THE BLAST FURNACE AND
THE MANUFACTURE
OF PIG IRON

An Elementary Trea-
tise for the use of the
Metallurgical Student
and the Furnaceman

By ROBERT FORSYTHE

SECOND EDITION

DAVID WILLIAMS COMPANY
11-16 PARK PLACE
NEW YORK

1909

To the Memory of

N. A. H.,

without whose aid

this volume might never have been attempted,

it is loyally dedicated.

PREFACE.

The author begs to say that this volume is not offered as an exhaustive treatment of the subject, but is designed primarily for beginners. He feels that recent writers on iron metallurgy have addressed themselves too exclusively to those who are already well versed in its mysteries. As a student, he sought in vain for a simple and concise statement of the general principles; as a teacher, he longed for one to recommend to his students; in practical work he has had to admit to many ambitious workmen that there was nothing in the literature of the subject that came within their grasp. For more than a decade he has felt that there was need of such a treatment on lines essentially American, and now has attempted to supply it. He is well aware that there are many others better fitted to undertake the task. Their failure to do so is the best possible reason for his presumption.

The Fahrenheit scale and the British Thermal Unit have been given preference in the text, because they are more frequently encountered in engineering writings in English, and because data in those units are more readily obtained in this country. The insertion of Centigrade equivalents is more confusing than helpful, and the change, if desired, is readily made.

The marginal references are intended primarily as guides to collateral reading. They may or may not indicate the source of authority for the contiguous statements. They may only refer to a discussion from another point of view. Chapter VII is not offered as a complete discussion of furnace design, but merely as a guide by which the untechnical furnaceman may test the suitability of his equipment. The chapter on the uses of pig iron is added in the belief that no manufacturer can approach his task intelligently unless he understands the limitations of his product. The brief review of the principles of chemistry and physics in Appendix I is intended primarily for untechnical readers, although it is hoped that it may serve to recall to others the principles which underlie the heat calculations in Chapter V.

The author wishes to acknowledge his indebtedness to all recent writers on ferrous metallurgy. In drawing from their writings, he has been compelled to condense the expressions, but he trusts that he has never perverted the sense. Besides thanking many manufacturers for information courteously furnished, he wishes to express his gratitude to Professor H. L. Smyth, of Harvard University, for a careful review of the chapter on Materials of Manufacture; to Mr. Alexander E. Outerbridge, Jr., of Philadelphia, for a thorough reading of and valuable suggestions on the parts relating to the constitution of iron and its use in making castings; to Mr. H. Clyde Snook, of the Roentgen Manufacturing Company, of Philadelphia, for a review of the thermal calculations and all the physical and chemical data; to Mr. F. W. Gay, Mechanical Engineer of the J. G. White Company, of New York, for a revision of the section on power development. Thanks are especially due to Mr. F. F. Amsden, Furnace Manager of the Central Iron and Steel Company, of Harrisburg, Pa., the value of whose many suggestions and patient revision of all parts pertaining to the design, construction and operation of the furnace it would be impossible to overstate.

R. F.

PHILADELPHIA, March, 1907.

ROBERT FORSYTHE.

Robert Foisythe was born at Braintree, Massachusetts, on September 5, 1869. He graduated from Harvard University in the class of 1894, and, after receiving his Master's degree in 1895, was for three years Instructor in Metallurgy at Harvard. He subsequently had practical metallurgical experience in the Open Hearth and Blast Furnace Departments of the Pennsylvania Steel Company at Steelton, Pa., and the Tidewater Steel Company at Chester, Pa. He died, after a short illness, on May 23, 1907.

At the time of the author's death, "The Blast Furnace and Pig Iron" was in proof. It is probable that he would have made various minor changes in the text. These it has been impossible to make, but it is believed that the book is published approximately as he wished. If any slight slips of pen or of calculation have been overlooked, the blame should rest, not upon the author, but upon the friend in whose hands were placed the proofreading and the final details of the book.   E. S

July 5, 1907.

TABLE OF CONTENTS.

Blast Furnace.

SUPPLEMENT.

Blast Furnace.

APPENDIX I.

THE BLAST FURNACE AND
THE MANUFACTURE
OF PIG IRON

Introductory.—Iron is the most abundant metal of this planet. In the form of some of its various compounds it is a constituent of practically all rocks and earths and is even found associated with many organic growths This wide distribution has much significance to the geologist and mineralogist, but the interest of the metallurgist is necessarily confined to those deposits which are sufficiently concentrated for the iron to be extracted at a profit Such concentrations are known as **ore bodies,** and they are the ultimate sources of all ferrous products

COMMERCIAL CLASSIFICATION OF IRON.

All of the iron of commerce may be classed under three general heads, namely: Cast iron, wrought iron and steel This distinction is based upon the physical characteristics of each class, as influenced by the presence in composition of certain non-ferrous elements. These three classes glide almost imperceptibly into each other through various intermediary forms which vary in accordance with the composition. It is well, therefore, to understand these basal forms in their original simplicity before proceeding to the consideration of the more complex.

The occurrence of non-ferrous elements in commercial iron is largely accidental. It is due to their incomplete elimination from the materials from which the iron is made The elements which are almost invariably present in all classes of ferrous products are five in number, namely, carbon, silicon, manganese, phosphorus and sulphur. In addition to these, rarer elements, such as arsenic, titanium, chromium, etc., are sometimes found. These elements occur alike in all classes of ferrous products, so that the above distinctions are based not on the character of the accessions but on the degree.

The iron of commerce is never pure. That form which usually contains the least foreign elements is **wrought iron.** Each element may comprise only a trace, or at most, only a few tenths of a per cent of the whole. The sum of all five may not reach one-half of one per cent. Such small quantities have little effect upon the properties of the metal, hence wrought iron may be said to exemplify the unalloyed element.

The peculiarities which distinguish wrought iron from cast iron and steel are as follows. It is softer and has a fibrous structure; it is extremely malleable and ductile, it has a higher melting point, and before the temperature reaches the melting point the metal passes through a plastic condition in which two pieces may be firmly united by the process of welding. The presence of foreign elements modifies all of these properties. It lowers the melting point, increases the hardness, imparts a crystalline structure, and interferes with the malleable properties and the power of welding

The ferrous product which most nearly resembles wrought iron is **steel.** Its composition differs from that of wrought iron in only one particular: it may contain more carbon. It may even be made from wrought iron by the addition of carbon alone—the other elements remaining unchanged. The amount of carbon which steel may contain usually ranges from a few hundredths per cent up to 2 per cent. or more. Its properties vary with its composition. When the content of carbon is low, it possesses the properties of wrought iron. It melts with difficulty; it is ductile and malleable, it is soft and may be welded. As the proportion of carbon increases, these properties pass through gradual changes. The melting point falls; the metal becomes harder and more resistant to pressure and changes of form, and the welding property gradually disappears. In addition, an entirely new property appears, namely, the power to become intensely hard when cooled suddenly from high temperatures. These characteristics become more marked with each addition of carbon The final product is a substance that is hard and brittle, and in every way different from the tough, fibrous, wrought iron

The third member of the series is **cast iron,** whose composition differs from that of the other two only in degree. The proportion of non-ferrous elements present in cast iron usually ranges from 5 to 10 per cent. The distribution of the more usual elements is as follows. Silicon, usually under 3 per cent., carbon, 2½ to 4½ per cent.; manganese, usually under 2 per cent.; phosphorus, usually under 1 per cent., and sulphur, usually under 0.1 per cent. Any or all of these elements except carbon may be reduced to a mere trace without declassifying the substance. It is evident, therefore, that the composition of cast iron, as well as that of steel, is variable; indeed, it is subject to a much wider variation. If we assume that all of the above elements be eliminated from a low carbon cast iron, we would have a substance that could be equally well classified as a very high carbon steel. A comparatively slight elimination of carbon would then produce a metal suitable for high grade steel instruments. Plainly, then, cast iron is but the completion of the series which was begun by wrought iron and continued by steel The continuousness of the composition and properties of this series is illustrated by the following diagram by Howe.

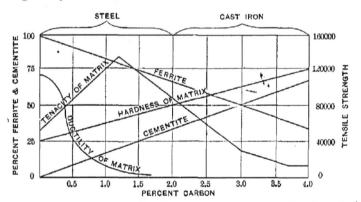

Am. Soc.
Test Mat.
II, p. 246.

Howe's Diagram Illustrating the Composition and the Properties of the Iron Series.

The properties of cast iron, with its accumulation of non-ferrous elements, could not be other than very different from those of the comparatively pure wrought iron. Since it is at the

other end of the series, its behavior is quite opposite to that of
wrought iron. It melts readily and passes suddenly into the
fluid state; hence, it cannot be forged and welded. Indeed, its
reluctance to change its form without rupture is accentuated
instead of decreased by high temperatures. It is coarsely crystalline, easily broken, and altogether different in appearance and
behavior from both of its associates.

. Cast iron in its **cruder** form, which is called " **pig iron**" from
a fancied resemblance, is the only ferrous product which is now
produced from the ores direct. It is a crude substance at best,
and while certain varieties may be formed directly into finished
products, such as castings, a very large proportion serves only
as a starting point for various processes of refining which result
ultimately in the production of wrought iron and steel.

CONSTITUTION OF PIG IRON.

Since pig iron, in its many forms, is made up of varying proportions of several different elements, each of which exerts certain specific effects upon the appearance and properties of the
metal, it may reasonably be considered a fairly complex substance.
Its complexity, however, does not depend solely upon its variable
composition, but is affected also by attendant circumstances. Initial temperature, rate of cooling, size, shape and position are
some of the conditions which will affect the properties of iron of
any given composition, and thereby multiply infinitely the varieties
possible   In order to grasp fully the possibilities of pig iron,
therefore, it is necessary to study its constitution in detail

CARBON.

As already stated, the only non-ferrous element which is absolutely essential to cast iron is carbon. Yet pig iron, which consists of iron and carbon alone, is of no practical value in making
castings. They would be hard, correspondingly brittle, and difficult to machine. Therefore other elements beside carbon are
necessary to the production of desirable castings. Their influence,
however, is due, not so much to their action upon the iron, as to
their effect upon the condition of the carbon.

Condition of Carbon.—As a rule the total per cent. of carbon in pig iron is fairly constant. It usually ranges from 3¼ to 4 per cent. of the whole. The variation in its condition, however, may cover a wide range. Carbon is known to exist in iron in no less than four different forms, only two of which it is necessary to consider here. These two forms are called graphitic carbon and combined carbon. **Graphitic carbon** is that which appears as black, shining flakes, readily distinguishable on a freshly-broken surface of high-grade pig iron. It is practically pure carbon, which has separated from the iron during solidification, and no longer exerts any influence upon the surrounding metal. The **combined carbon** is that portion of the total carbon which still remains in combination, or alloyed with, the iron which surrounds the flakes of graphite. It exerts a profound influence upon the metal.

In general, pig iron may be likened to a mass of concrete, in that it consists of particles of graphite surrounded by a matrix of metallic alloy. The proportion of carbon existing in each state is subject to wide variation. In some irons it may exist as practically all graphite with very little combined carbon; in others it may be all in the combined form with little or no graphite. Its condition is profoundly affected by the presence of other elements and also by the rate of cooling. Very important distinctions in pig iron are based upon the ratio of graphitic and combined carbon. The presence of graphite gives the grayish black crystalline appearance which is characteristic of foundry irons. Much combined carbon, on the other hand, gives iron a white or mottled appearance and renders it unsuitable for many purposes.

Structure of Pig Iron.—The strength of iron is greatly affected by the condition of the carbon. The crystals of graphite are brittle and show decided cleavages, hence they cannot be a factor of strength in the iron. Indeed, by breaking up the continuity of the metallic matrix, the graphite causes weakness, which will vary directly with the quantity. The matrix is an alloy of iron with carbon and other elements which may be present. It possesses properties varying with the percentage of carbon in the alloy. Since all of the strength of the iron is due to the metallic

Tr. A. I. M
XXXI p

matrix, it is evident that the strength will vary in proportion to
the amount of carbon entering into the matrix. It appears, then,
that carbon affects the strength of iron in two ways: (1) in the
form of graphite it weakens, by interrupting the continuity of the
matrix, (2) as combined carbon it strengthens by strengthening
the matrix. This is true, however, only up to about 1 per cent.
of combined carbon. Further quantities tend to weaken the
matrix. The explanation of this reversal of the effect of carbon
lies in the structure of the alloy. Under the microscope, combinations of iron and carbon, which are formed under usual conditions, show a crystalline structure which may be likened to that
of granite, and which consists of two components. All of the
carbon unites with iron in the ratio of 1:14 by weight, to form a
carbide of iron having 6⅔ per cent. carbon and 93⅓ per cent.
M. E.,
p 734. iron, and approximating the chemical formula Fe_3C, which is
called **Cementite.** This is one component. The other consists of
the free iron which remains in excess of the quantity required to
form Fe_3C with all of the carbon present, and is known as **Ferrite.** The ferrite and cementite then proceed to unite in alternate
parallel layers to form a true eutectic alloy, in which they approximate the proportions 7:1 by weight. Under the microscope this
finely-banded structure presents a pearly luster, and is consequently known as **Pearlite.** The pearlite crystallizes in tolerably
regular polygonal grains, along whose edges the excess of ferrite
or cementite collects as a network. Since cementite contains
6⅔ per cent. of carbon, and unites to form pearlite with seven
times its weight of ferrite, it follows that pearlite must contain
$\frac{6⅔}{7+1} = 0.85$ per cent. of carbon. It is apparent, then, that the
presence of just 0.85 per cent. of carbon would produce a metal
consisting entirely of pearlite, and it is a necessary corollary,
therefore, that a metal containing less than 0.85 per cent. would
consist of pearlite and a residual network of ferrite, while one
containing more than 0.85 per cent. of carbon would consist of
pearlite and a residual network of cementite. Since pearlite is
a very strong substance, the natural planes of weakness in such
a metal would be along the lines of network. Ferrite, being pure
iron and naturally tough in consequence, would offer more re-

sistance to cleavage than cementite, which is a hard crystalline substance. This fact explains why a matrix containing less than 1 per cent. combined carbon is stronger than one containing more than 1 per cent. The proportions of ferrite, cementite and pearlite, and the excess of ferrite or cementite over that needed to form pearlite with all of the carbon present, are shown by the following table for various percentages of carbon:

Carbon.	Ferrite	Cementite	Pearlite.	Excess ferrite	Excess cementite.
0.0	100 0	0.0	0 0	100.0	0.0
0 1	98 5	1 5	12 0	88.0	0.0
0.2	97 0	3.0	25 0	75 0	0.0
0 3	95.5	4 5	37.0	63.0	0 0
0 5	92.5	7.5	62 0	38.0	0 0
0 7	89.5	10.5	87.0	13 0	0.0
0 85	87 8	12.2	100 0	0.0	0.0
1 0	85.0	15 0	97 0	0.0	3.0
1.5	77 5	22 5	88 0	0.0	12 0
2.0	70 0	30.0	80 0	0.0	20.0
2 5	62.5	37.5	71.0	0 0	29 0
3.0	55 0	45.0	62 5	0.0	37.5
3 5	47 5	52 5	54 0	0 0	46 0
4.0	36 25	63.75	45 5	0.0	54 5

Am. Soc. Test. Mat II, p 246.

Strengthening Effect of Carbon.—From these considerations it appears that a change in the condition of carbon may sometimes strengthen and sometimes weaken the iron. For example, an iron containing 3½ per cent. graphite and ¼ per cent. combined carbon would have a highly interrupted matrix which contains but little of the strengthening pearlite. Any change of graphite to combined carbon would make the matrix more continuous and at the same time increase the pearlite. Both factors contribute strength. The maximum strength would probably be reached when there was 2¾ per cent. graphite and 1 per cent. combined carbon. A further change of graphite to combined carbon would tend to increase continuity of matrix, but would introduce weakness into the matrix itself, in the shape of brittle cementite. The weakening effect of cementite overbalances the strengthening effect of increased continuity and the net result is a loss of strength. Further change in this direction only tends to aggravate the weakness through the introduction of still larger masses of the brittle cementite. These effects may be masked, however, by other conditions or the presence of other elements, but it is very evident that the strongest iron, as far as carbon is concerned, will be the one which contains about 1 per cent. of combined carbon, and the least possible graphite. This statement is well borne out by experience and is well illustrated by analyses selected from experiments by Johnson:

Si.	S.	P.	Total C.	Graph. C.	Comb. C.	T S , lbs. per sq. in.	Character of fracture.
1.20	0.066	0.175	3.87	3.72	0 15	17,500	Very open grain.
1 21	0.067	0.179	3 85	3 52	0 33	22,400	Bright open grain.
1.20	0.006	0.170	3 82	3.42	0.40	24,550	Bright close grain.
1 21	0 061	0.174	3.88	2.95	0.93	33,850	Very close grain.

Since these irons are remarkably uniform in composition, we can attribute their change in strength and texture only to the change in condition of the carbon.

Absorption of Carbon.—Carbon may be taken up by iron in two ways. It is very readily dissolved by molten iron, and it is also more or less actively absorbed by the solid metal at all temperatures above redness. The dissolving of carbon by molten iron up to the saturation point is practically instantaneous. On the other hand, the rate of absorption of the solid iron and its capacity for carbon depend upon the temperature. The quantity absorbed at a given temperature depends also upon the time allowed. The actions of these two methods of uniting carbon and iron are well illustrated by experiments by E. H. Saniter. He melted some pure iron wire in a crucible with an excess of carbon and found that the iron absorbed 4.73 per cent. of its weight of carbon. He heated some of the same wire in a porcelain tube which was packed full of powdered charcoal, and held it at a temperature of 1650 degrees F. At the end of 7 hours it contained 1.64 per cent. C., at the end of 14 hours it contained 2.79 per cent. C., at the end of 21 hours it contained 2.95 per cent. C., which was probably reasonably close to the saturation point for that temperature.

Carbon and Other Elements.—The quantity of carbon that may be dissolved by molten iron, and the state in which the dissolved carbon will exist after cooling, are both affected profoundly by the presence of other elements. Silicon, for example, changes combined carbon to graphitic. It also tends to exclude carbon.

An iron containing 3 per cent. silicon usually contains 3 to 4 per cent. carbon. With 10 per cent. silicon, not more than 2 per cent. carbon may be present, and with 20 per cent. silicon, the carbon is nearly or wholly absent. On the other hand, manganese and chromium raise the saturation point of iron for carbon, and tend also to keep it in the combined state. The presence of 75 per cent. manganese permits the solution of about 7 per cent. carbon, while a similar proportion of chromium will raise the saturation point above 10 per cent These figures represent, however, more nearly the dissolving power of manganese and chromium than that of the iron, as the percentage of iron in such metals is comparatively insignificant. In irons containing both silicon and manganese, such as the silico-spiegel series, these elements appear to share the honors. Silicon continues to exclude carbon, though not as rapidly as when alone, but the ratio of combined carbon to graphite is higher than in ordinary irons. The following analyses by W. J. Keep illustrate the effect of silicon upon the quantity and condition of carbon in cast iron:

Si	0.18	1.25	2.08	3.15	4 89	5.89	9.10	10 34	12 08	16 27	20.00
T. C	2.53	3 55	3.75	3 50	3 44	3.15	2 58	1.99	1.58	0 75	...
G. C	0 49	3.22	3.12	2.46	3 40	2 85	0.90	1 92	1.52	. .	...
P	0.26	0.08	1.65	1 06	1.42	1.10	0.09	0.48	0.48	0 01	...
S	0 03	0.04	0.01	0.02	Tr.	0.02	0.03	Tr	Tr.	0.01	...
Mn	0 00	0 18	0 87	1 35	...	1 00	2 20	0 57	0 76	0 60	...

. SILICON.

Silicon appears to be able to unite with iron in all proportions. Pig irons containing quantities of silicon ranging from a few tenths per cent. up to 5 per cent. are made, at least occasionally, at all blast furnaces. Ferro-silicon containing 12 per cent. silicon is a regular blast furnace product. By means of the electric furnace the percentage may be raised to 60 or 80 per cent.

Effects of Silicon.—The effect upon iron of small percentages of silicon is usually masked by that of carbon. In the presence of high percentages, however, carbon is practically excluded and the resulting metal is hard and brittle It is probable, then, that silicon, even in small quantities, is really a hardening agent, but owing to its effect upon the condition of carbon, usually appears to be a softener. Analyses by Johnson appear to indicate that when other conditions are constant, silicon materially weakens iron:

Inst Jour,
1898, II.,
Table III

T. C.	G. C	C. C	P.	S	Si.	T S., lbs. per sq. in	Character of fracture.
3.82	3.03	0.79	0.179	0.071	0.31	31,550	Gray fracture
3.53	2.98	0.85	0.178	0.070	1.27	29,050	Gray fracture.
3.90	3.07	0.83	0.179	0.067	1.76	22,850	Gray fracture.
3.81	3.01	0.80	0.179	0.073	2.03	19,650	Gray fracture

Even in small percentages silicon appears to exclude carbon from iron. Pure iron has a capacity for carbon up to 4 7 per cent. of its own weight when melted at the highest temperatures. Irons containing 2 to 3 per cent. silicon and made at normal furnace temperature rarely contain over 3.5 to 4 per cent. of carbon, which shows the exclusion of nearly or quite 1 per cent. of carbon. Low silicon irons also rarely show over 4 per cent. of carbon. This fact may be accounted for by the low temperature of formation of such irons.

In quantities exceeding about 4 per cent., silicon gives to pig iron a peculiar glazed appearance, accompanied by increasing weakness, which unfits it for castings. The silicon in foundry irons does not usually rise above 3 per cent., and for most purposes 2 per cent is better No exact rule can be given for determining the best percentage of silicon for given results, because its influence depends upon its power to control the *ratio* of graphitic and combined carbon, while the strength of the iron, as we have seen, depends mostly upon the *quantity* of combined carbon. Since silicon affects only the *proportion* of the total carbon which remains in combination with the iron, it is evident that the *quantity* so uniting will depend upon the total amount of carbon present. This quantity is materially affected not only by the silicon present, but also by the temperature of formation and the presence of other elements. It is evident, then, that the proper quantity of silicon for certain results cannot be easily predicted, but that it varies according to the percentages of other elements present.

Tr A I M E,
XXVIII,
p 789

Benefits of Silicon.—If a casting were made from pig iron practically free from silicon it would be found to be hard, white, brittle and more or less porous. The addition of a small percentage of silicon would make the casting solid, and thereby add to its strength. Further addition of silicon, say 1 per cent., would turn the iron from white to a fine-grained gray color, and increase the softness and strength. Still further additions, up to 3 per cent., would enlarge the grain and increase the softness without materially decreasing the strength. Silicon up to 0.9 per cent. usually increases the strength of all sizes of castings. When it exceeds 1 per cent. it tends to decrease the strength of all sections over 1 inch square. Up to 2 per cent. it strengthens all sections under 1 inch square, and up to 3 per cent. it strengthens sections of ½ inch square. Above 5 per cent. of silicon the color becomes lighter, the hardness returns rapidly, and the strength is soon changed to brittleness. In general, it may be stated that up to about 3 per cent., silicon is increasingly beneficial. since it promotes soundness, softness and strength, and decreases shrinkage and chilling properties. These effects are the result of changing combined carbon to graphite, in accordance with the law that combined carbon usually varies inversely with the quantity of silicon. This effect is not always proportionate, however, as iron appears to be more sensitive to change in silicon when it is below 1 per cent. than when above. By the addition of silicon to white iron when molten, practically worthless iron may be turned to good account. However, carbon has a tendency to retain its original condition, so it may happen that a given percentage of added silicon may not produce the same result as if the silicon had been originally present in the pig. When enough silicon has been added to give soundness, even while the iron is still hard and brittle, it offers its highest resistance to compressive forces. At that time, also, it exhibits its maximum specific gravity. When the addition of silicon begins to cause the separation of graphite, the transverse strength reaches its maximum. The greatest tensile strength is reached when the silicon rises toward 2 per cent. Deflection follows strength closely, and shrinkage follows hardness. The ideal compositions recommended by Turner are:

	Per cent. Si.
Maximum hardness, under	0 80
Maximum crushing strength, about	0 80
Maximum transverse strength, about	1 40
Maximum tensile strength, about	1 80
Maximum softness, about	2 50

In addition to its beneficial effect upon carbon, silicon improves cast iron in two ways; it tends to exclude sulphur and it prevents blow holes. Irons which are high in silicon are generally low in sulphur. This is probably largely due to the volatilization of the sulphur by the high temperature necessary to produce high silicon irons. The action of silicon in promoting soundness by preventing blow holes is not clearly understood Blow holes are caused by the gases which are always in solution in the molten iron and which, set free during the cooling, become entangled in the rapidly solidifying metal. The presence of the silicon either prevents the dissolving of the gases in the first place or else it enables them to stay in solution in the solid metal. The latter explanation is the one usually accepted.

MANGANESE.

Manganese can alloy with iron in all proportions. Up to about 1 per cent. it does not appreciably affect the properties of cast iron. In larger proportions it increases the hardness of the iron and also its capacity for carbon Unlike silicon, however, it tends to keep the carbon in the combined state. Therefore it can harden the iron in two ways: by its own effect, and by its influence upon the condition of the carbon. As the proportion of manganese rises, hardness and brittleness increase. Between 10 and 30 per cent. manganese gives a hard, white, brittle substance, which shows large crystal facets on the fractured surface. This is known as spiegeleisen, and is made regularly for use in steel making. It is wholly unsuitable for making castings. Irons containing higher percentages of manganese are known as ferromanganese. They usually contain as high as 80 per cent. Mn, and, like spiegel, are used chiefly in steel making. Their grain is close and the fracture granular, frequently showing beautiful iridescent colors. They are so brittle as to be easily pulverized.

Effects of Manganese.—In small quantities manganese is undoubtedly beneficial to cast iron. Less than 1 per cent. does not cause a material increase in hardness. Indeed, it may act as a softener of certain kinds of hard iron. This is due to its power of uniting with sulphur to form a sulphide of manganese which separates from the metal. Hence, an iron whose hardness is due to the presence of sulphur may be softened by the addition of manganese. It is therefore particularly beneficial to irons which do not have much silicon, as it keeps the sulphur low. Its presence tends also to prevent the absorption of sulphur from the fuel during remelting. The affinity of manganese for sulphur forms the basis of a process of desulphurization of molten pig iron. Inst. Jour., 1891, II, p. 70.

In quantity exceeding 1 per cent., manganese is likely to be detrimental to cast iron, partly because of its tendency to increase the percentage of carbon, and to keep it in the combined state, thereby tending to make the iron white, and partly because of its own power to confer hardness. Hardness, from any cause, is followed by a train of evils, such as brittleness, increased shrinkage and difficulty in machining. In remelting, manganese is readily oxidized and lost from the iron, so that the content of the pig cannot be safely counted on in the castings. The effect of manganese on the condition of carbon, the quantity of sulphur, and chill is illustrated below: Inst. Jour., 1898, II Iron Age, June 29, 19(p. 205.

Si.	P	T. C	Mn.	C. C.	S.	Chill
0.55	0.22	3.67	1.38	0.62	0.048	2.00 inches.
0.49	0.25	3.72	2.00	0.74	0.030	2.25 inches.
0.49	0.24	3.70	2.25	1.03	0.025	2.30 inches.
0.74	0.27	...	3.80	2.52	0.020	All white.

PHOSPHORUS.

Phosphorus and Carbon —Phosphorus appears to be able to unite with iron in all proportions. It exercises a strong excluding effect upon carbon. Iron in combination with 15.58 per cent. phosphorus forms a compound corresponding to the formula Fe_3P, in which carbon is absolutely insoluble. It is evident, then, that an iron containing about 16 per cent. phosphorus will be carbonless, and that lesser quantities of phosphorus will admit of only a proportionate amount of carbon. This is shown by the following analyses by Stead. Inst. Jour., 1900, II., p 109

Si.	P.	T. C.	G. C.	C. C.	Mn.
0.92	0.21	3.72	2.62	1.10	Trace.
1.96	4.95	2.29	1.73	0.56	Trace.
1.96	6.85	1.99	1.88	0.11	Trace.
2.84	8.35	1.69	1.69	0.00	Trace.
3.36	12.86	0.83	0.83	0.00	Trace.

Since the ratios between the graphite and the combined carbon in this table are about what would be predicted from the quantity of silicon present, it is probable that phosphorus does not exert a very strong influence upon the condition of the carbon, but only upon its quantity.

Phosphorus tends strongly to segregate, and high phosphorus pig when broken before the center has solidified shows generally several times as much phosphorus in the liquid portion as in the solidified portion. This liquate is probably an eutectic alloy of the elements present. The eutectic alloy of iron and phosphorus begins to appear in pig iron when phosphorus exceeds about 1.7 per cent. Below this point the phosphorus exists as the compound Fe_3P, which is dissolved in the excess of iron. It is this solution that the presence of carbon tends to prevent. The iron does not appear to be able to hold in solution both carbon and Fe_3P independently. One must always be sacrificed for the other. This struggle for the possession of the iron is illustrated by the following analyses by Stead

C.	Total P.	P in Fe_3P, dissolved in iron.	P in free Fe_3P.
0 00	1 75	1.75	0 00
0 18	1.77	1.18	0.59
0.70	1 75	0 75	1.00
1 40	1 76	0 60	1 16
3 50	1.71	0.31	1.40

Effect on Strength.—The Fe_3P, thus set free, forms an eutectic with the other elements not in solution, which then combines with the pearlite to form a phospho-pearlite eutectic containing 0.6 per cent. P. Any excess of Fe_3P collects along the junctions of the grains and causes brittleness in high phosphorus irons. This tendency is greatly lessened by the fact that the phosphorus compounds tend to collect in globular form instead of completely enveloping the grains. This distribution leaves considerable areas of strong metal between, and postpones the appearance of the weakness until the amount of phosphorus causes such large globules that the strong ground-mass is materially lessened. This condition is reached at approximately 1.7 per cent. The weakening effect of phosphorus is largely neutralized by the presence of titanium. The titanium increases the total carbon and makes it mostly graphitic. It is claimed that ½ per cent. of titanium will give strength to iron containing 3 per cent phosphorus   Phosphorus causes pig iron to have an enlarged grain with a yellowish color. It does not affect the shape of the grain, but only the size, which is additional reason for weakness.

Benefits of Phosphorus.—Phosphorus may be beneficial on account of its tendency to eliminate blow holes, thereby promoting soundness. It decreases shrinkage also, but does not materially affect the hardness. Its most distinctive characteristic is its power to render the iron more fluid, or to prolong the period of fluidity to such an extent that the metal has opportunity to penetrate to all parts of the mould, thereby producing sharp impressions. This quality commends it especially as a constituent of irons to be used in making castings of fine, intricate patterns which do not require much strength. In castings that need strength, the phosphorus should not be allowed to rise much above ½ per cent. Where fluidity is the prime requisite 1 per cent. or even more may be desirable.

Tr. A. I. M. E.
XXVI., p 144

Tr A. I. M. E.
XVIII., p. 455.

SULPHUR.

In general, it may be stated that from a metallurgical standpoint sulphur in iron presents no redeeming features. Therefore it is generally allowed to be present only in quantities which are small compared to those of other elements. Its presence is vigorously opposed by manganese, and also by the furnace conditions that produce high silicon. As would be expected, therefore, it is usually low in high-silicon irons   High sulphur is usually an accompaniment of low-silicon, because the low temperature of formation permits the absorption of sulphur.

Effect of Sulphur.—Since silicon is usually low, high sulphur is generally accompanied by much combined carbon, and therefore it has acquired the reputation of being a hardener of iron. It seems to be clearly established that the presence of sulphur is accompanied by an increase in hardness and shrinkage of castings, together with increased density and liability to crack. Its presence may not be injurious in certain classes of work, but for castings which must be accurate to the pattern, or must be machined, sulphur should not rise much above 0.05 per cent. When manganese is

Tr A. I. M. E
XXIII., p 382.

present in quantity above ½ per cent. the greater part of the sulphur is found combined with it as manganese sulphide. In this condition it does not exert its usual deleterious effect upon the metal, and if the period of fluidity is sufficiently long this sulphide will float up to the surface and separate itself entirely from the metal.

SUMMARY OF INFLUENCE OF METALLOIDS.

The effect of the metalloids upon cast iron may be summarized as in the following table. By this it may be seen that the effects of manganese and sulphur are somewhat similar, while those of silicon and phosphorus are opposed to them:

	Si.	Mn.	P.	S.
Total C.	Decreases.	Increases.	Decreases.	Neutral.
Graphite.	Increases.	Decreases.	Neutral.	Decreases.
Comb. C.	Decreases.	Increases.	Neutral.	Increases.
Soundness.	Promotes.	Promotes.	Promotes.	Neutral.
Strength.	Decreases.	Decreases.	Decreases.	Decreases.
Shrinkage.	Decreases.	Increases.	Neutral.	Increases.
Chill.	Decreases.	Increases.	Neutral.	Increases.
Hardness.	Decreases.	Increases.	Neutral.	Increases.
Fluidity.	Decreases.	Neutral.	Increases.	Neutral.
Sulphur.	Excludes.	Excludes.	Neutral.	....

The diagram below illustrates graphically the opposing effects of silicon and manganese upon the other three elements.

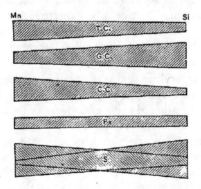

Diagram Illustrating Effects of Silicon and Manganese.

ALUMINUM.

Aluminum cannot be looked upon as a constituent of pig iron, but its cheapness has made it available of late years as a metallurgical reagent. It may be added, therefore, whenever the effects which it produces are desired. In general its effects resemble closely those of silicon. It changes combined carbon to graphite, it increases strength and decreases shrinkage, chill and fluidity. Its effect is most marked between 0.5 per cent. and 2 per cent. Above 2 per cent. there is a decrease of total carbon, and above 5 per cent. its power to form graphite is practically gone. It never decreases the combined carbon below 1 per cent., which indicates that its effect is not so complete as that of silicon, although it is more active in small quantities. Between 0.5 per cent. and 4 per cent., chilling appears impossible. Its influence is well shown by the analyses of Melland and Waldron in the following table:

Tr. A. I. M. E., XVIII, p. 102.

Inst. Jour, 1900, II., p. 244.

Si	0.24	0.23	0.21	0.19	0.21	0.28	0.22	...	..	0.26
Al	0.00	0.02	0.16	0.25	0.53	1.78	3.82	4.24	8.31	11.80
Total C.	3.90	3.93	4.00	3.96	3.83	4.07	3.59	3.57	3.32	3.12
Graph. (chilled)	0.38	0.30	0.32	0.91	3.06	2.96	2.53	2.28	0.66	0.20
Graph. (sand)	0.38	1.20	3.01	3.49	2.93	2.93	2.54	2.49	0.99	0.20

OTHER ELEMENTS.

Occasionally rarer elements are found in pig iron. Almost all ores contain traces of **arsenic,** which is readily reduced and enters the iron. It forms with it an arsenide, which is weak and brittle; therefore its presence has an injurious effect. As it rarely occurs in quantity exceeding 0.04 per cent., however, the effect is hardly noticeable.

Inst. Jour, 1888, I, p. 171.

Titanium is reduced with difficulty in the blast furnace, and rarely occurs in pig iron in quantity exceeding 1 per cent. In small quantities it appears to have a very beneficial effect. It increases the density of the metal, giving it greater tenacity and resistance to wear. When chilled, the metal resists the hardest steel and is especially good for car wheels. Wrought iron or 'steel made from titaniferous pig are unusually strong and tough and never show red-shortness or cold-shortness. Whether the beneficial effect is due to the small traces of titanium or to

Iron Age, Dec. 11, 1902, p. 22.

Tr. A I M E,
XXI , p. 832 the freedom from phosphorus and sulphur, which is the peculiarity
of all titaniferous ores, is not clearly proven. Iron made from
titaniferous ores without absorbing any titanium appears to be
superior to ordinary irons.

PHYSICAL PROPERTIES OF CAST IRON.

The effect of various physical conditions upon the properties
of cast iron may be as great as the effects of variation in composition itself. The two are so inextricably bound together that
it is generally impossible to say where one ends and the other
begins.

INFLUENCE OF HEAT ON CAST IRON.

Specific Gravity.—The matrix of cast iron tends to crystallize in octahedra, surrounding flakes of graphite. The presence of
the graphite lowers the specific gravity of the metal in the ratio
of 2 2 to 7 8. The graphite represents 12 to 15 per cent. by bulk,
and as a result, gray cast iron usually has a specific gravity of about
7.2. The specific gravity of the metal is greatly affected by the
condition of the carbon and is therefore variable. Iron which has
a specific gravity of 7.25 when gray, may have over 8 when
chilled. This is due to the contraction of the metal during cooling. Chilled samples generally show considerable shrinkage spaces
in the top, while gray iron, on the other hand, when made to
solidify between rigid walls, often extrudes some metal during
solidification. The rate of cooling also affects the specific gravity,
so that small sections always have a lower specific gravity than
large ones cast from the same metal. The specific gravity of cast
iron when molten is less than when cold, but greater than when
Iron Age,
Aug 10, 1905,
p 348. hot. Solid gray iron when thrown into a bath of molten iron
will sink at first, but when heated through will rise and float until
melted. The reason for these phenomena is revealed by observation of the curves made by cast iron when cooling

Expansion and Shrinkage.—If a test bar be cast between a
fixed surface and a movable contact which is connected to a recording device, it will show a series of expansions immediately
after becoming solid. Gray iron shows three distinct expansions

which are scarcely distinguishable in white iron Each expansion probably represents a period of molecular rearrangement. Inst Jour., 1895, I, p 227 The third expansion, which is the most marked, is due to the separation of graphite and is followed by regular contraction until the iron is quite cold. The net shrinkage of the casting will be the algebraic sum of these expansions and contractions, and therefore will be smaller the greater the expansion. Since the expansion is caused by the separation of graphite, it follows that shrinkage will be decreased by any condition which creates graphite, such as the presence of silicon, large cross section, slow cooling, etc. For example, a cross section of ½ inch with 2½ per cent. silicon shows no more expansion than a 2-inch section with 1 per cent. silicon. A 1-inch section will show a maximum expansion inside of 20 minutes, while a 4-inch section needs at least an hour. Expansion is decreased by the presence of sulphur, but not appreciably affected by phosphorus and manganese. The temperature at which the separation of graphite occurs has been determined to be between 1300 degrees F. and 1400 degrees F. The other two expansions, when present, occur at about 1650 degrees F. and 2000 degrees F. They are all characterized by a Iron Age, May 24, 1906, p 1671 retardation in the rate of cooling, which indicates an evolution of heat resulting from the molecular changes.

Melting Point of Cast Iron.—Moldenke has found that the Iron T R , Oct. 27, 1898. melting points of pig irons range from close to 2000 degrees F. for white irons, to nearly 2300 degrees F. for gray. That graphite recombines with iron before melting is proven by the fact that gray iron heated to whiteness but not melted shows the third expansion on cooling again. The higher melting points of graphitic irons are due to the delay necessary to effect this recombination. The variation in the apparent melting points is not strictly in proportion to the quantity of graphite present, because the temperature necessary is greatly influenced by the rate of reabsorption of graphite

Pipes and Cavities.—When molten iron is poured into a mould, the surface of the metal which is in contact with the relatively cold mould solidifies first and expands during the first few minutes of cooling. By the time it starts to contract, it finds itself opposed by the expansive force of some subsequently cooled

material. As a result, before contraction can take place, the enveloping shell of metal becomes so rigid that it cannot shrink
to accommodate itself to the contracting interior, and unless fresh
molten metal is supplied by means of " risers " or " sinkheads,"
pipes or cavities may result.

Rate of Cooling —The prolongation of fluidity and the different rates of cooling tend to foster two undesirable conditions, viz ,
segregation and internal strains. By **segregation** is meant the
partial concentration of one or more of the metalloids in any
portion of the mass of metal   Such concentrations must occur
before solidification and, as they take time, are naturally more
marked in the centre of the mass, since it remains fluid longest.
The condition is fostered, also, by the presence of those elements
which lower the melting point of metallic iron, particularly carbon and phosphorus. Uniformity of composition is promoted by
shortening the period of fluidity as much as is practicable. Sudden cooling, on the other hand, sets up **internal strains,** which
may cause more damage than segregation. It is not uncommon.
for castings to break quite unaided on account of internal strains
Such a condition may be relieved by **annealing,** that is, by allowing the mass of metal to cool slowly from a high temperature,
thereby giving time for molecular adjustment in the solid condition. The annealing temperature should be uniform and well
above redness. The temperature may be due to the initial heat
of the casting, or to a subsequent reheating. Annealing usually
decreases the strength of cast iron by changing some of the
combined carbon to graphite, but it usually strengthens a casting
through permitting the adjustment of initial internal strains.

Effect of Shock.—It was first observed by Outerbridge that
an effect similar to annealing could be produced in castings by
successive shocks, such as result from cleaning them in a tumble-
barrel or tapping with a hammer   He attributed this result to
the mobility of molecules of iron which appear able to readjust
themselves even in the solid state. Keep observed that similar
treatment also makes slight changes in the volume and strength
of castings.

fr A I M E,
XXVI , p 176.

Ibid,
XIX, p 351

Effect of Repeated Heatings.—That cast iron undergoes a
marked change in volume through the action of heat was proved

by Outerbridge. Repeated heatings and coolings result in permanent expansion, at times causing serious buckling. The change in volume is due to continuous pushing out of the crystals, which do not return to place on cooling, but leave open spaces. The effect is most active at 1450 degrees F., at which temperature 33 treatments are required to exhaust the action. A bar 1 inch square and about 15 inches long gained 12 per cent. in length and 14 per cent. in thickness, giving a total increase in volume of more than 40 per cent. The specific gravity fell below 5, and the strength to about one-third of that which it originally possessed.

Iron:
Feb.:

STRENGTH OF CAST IRON.

When pig iron is to be converted into steel or wrought iron, the knowledge of its physical properties is of little value. It is important only that the composition of the pig should be appropriate to the method of conversion. If, however, the iron is to be used as a material of construction, in the form of castings, its physical properties are of vital importance. The property which naturally demands first consideration is *strength*, in as much as a weak metal would be of comparatively little use in structures. Of scarcely less importance, however, is *hardness,* or, more properly, softness, since on account of their natural roughness most castings need to be smoothed or shaped in places in order to make accurate contacts. Excessive hardness, therefore, should be avoided when much machining is necessary.

TESTING CAST IRON.

The most usual tests for determining the strength of cast iron are tensile, compressive and transverse.

Tensile Strength.—The tensile tests are made by pulling a test bar of carefully measured cross-section until it breaks. The strength needed to produce rupture is known as the ultimate tensile strength and is usually expressed in pounds per square inch of cross-section. The tensile strength of cast iron is low as compared to that of wrought iron and steel, hence it is almost never used for tensile purposes. This test is, therefore, of little importance commercially, although it furnishes a comparative guide to mechanical properties.

Compressive Strength.—Compressive tests are usually performed by submitting small cubes of the metal to a crushing force until rupture takes place. The results are usually expressed in pounds per square inch of cross-section of the metal. The resistance to compression of cast iron is so high that tests are usually considered unnecessary.

Transverse Strength.—Transverse strength tests are applied to the middle of a test bar which is supported at the ends. They are of two kinds, known respectively as "dead-load" and "impact" tests. A **dead load** test is a static test, applied by quietly increasing the gravitational force on the bar without sudden shocks. The **impact** test is a dynamic test applied by means of the momentum of a known weight falling from a known height under the force of gravity. The results in each case are usually expressed in terms of pounds per square inch of cross-section of the test bar. Transverse tests are the most important, hence the most usual

Testing Terms.—The amount of force applied in making strength tests is usually known as the **stress.** The effects of the stress upon the test piece are called **strain.** The amount which a given stress on a straight bar will make it deviate from a straight line is called **deflection,** and is measured usually in fractions of an inch. If the deflection is completely obliterated on the removal of the stress, the bar is said to be **perfectly elastic.** Most metals are perfectly elastic up to a certain limiting stress where elasticity ceases. This stress is the measure of the **elastic limit** of the metal. When a stress in excess of the elastic limit of a metal is applied and removed, the metal does not conform to its original shape but retains a greater or less deflection, which is known as **permanent set.** This is measured in the same way as deflection. A body which shows no deflection under stress is said to possess **perfect rigidity.**

Testing Conditions.—Since cast iron is subject to such wide variation in composition, and its properties are so profoundly affected by attendant circumstances, it follows that in order that

Keep's
Cast Iron."

tests may be reasonably comparable, the test bars should be made similar in all respects The following are some of the causes of errors: Different sized test bars cool at different rates and therefore the larger the test bar the weaker it will be per unit of cross-section. Square bars show greater strength for the same cross-section than round ones. Bars cast horizontally are stronger than those cast vertically, and those cast in green sand show more strength than those cast in dry sand. Bars from a large casting will appear stronger than those from a small one. The centre of a casting is always weaker than the outside. Low silicon gives strong large castings and weak small ones, while high silicon gives strong small castings and weak large ones.

Am Soc
Materia
III., p :

The standard test bar for cast iron is 15 inches long by 1¼ inches square, moulded in damp sand, from which the pattern is withdrawn without tapping The bars are cleaned by brushing only, and the load applied gradually midway between supports, 12 inches apart.

CHAPTER I.

MATERIALS OF MANUFACTURE.

As pointed out in the introductory chapter, cast iron in the form of pig iron is practically the only ferrous product that is now produced direct from iron ores. The operation is carried out in a tall, cylindrical furnace, known as a "blast furnace," into the top of which the raw materials are charged, and from the bottom of which the iron is drawn in the molten state.

The materials which go to make up the charge of a blast furnace fall naturally under three heads; namely, ores, fuels and fluxes.

The ores comprise the iron-bearing portion of the charge The iron may be present in the ore in any of its many compounds, providing that the ore is rich enough to pay for the treatment The compounds most usually treated are oxides.

The fuels are necessary in order to furnish the heat needed to carry on the reactions which take place in the furnace, and to melt the resulting products. They furnish, also, the reagent for removing the oxygen from combination with the iron.

The duty of the fluxes is to make fluid the infusible earths which accompany the charge, by uniting with them to form combinations that may be readily melted and drained from the furnace

ORES OF IRON.

An ore of iron consists primarily of two constituents; viz., a compound of iron mixed with a gangue of earthy materials. It may be compared, for purposes of illustration, to a mixture of iron-rust and dirt. The problem of the smelter is the extraction of the iron from its combination, and the fluxing of the earthy constituents When these reactions take place at sufficiently high temperatures, we have, as products of the operation: (1) the molten metal which has been separated from its combination and melted, and (2) a liquid slag, which is a new compound formed by uniting the flux with the earthy materials which constituted the gangue. These two products have very different specific weights, and if allowed to lie quietly, separate into two layers, the lower, pig iron, the upper, slag. By means of tapping-holes at different levels, they may be drawn off separately.

CONSTITUTION OF IRON ORES.

The compounds of iron best suited to smelting in a blast furnace are the oxides, although carbonates and silicates are sometimes used. Oxides of iron are compounds of iron and oxygen which have the formulas Fe_2O_8 and Fe_3O_4, and contain respectively 70 per cent. Fe and 72.4 per cent. Fe. The carbonate has the formula $FeCO_3$ and contains 48.3 per cent. Fe. The silicate may be represented by the formula $FeSiO_3$, which contains 42.4 per cent. Fe. It is evident, then, that from the point of richness, oxides are far superior to carbonates and silicates. Moreover, the oxides yield metallic iron upon the simple removal of the oxygen, whereas in the cases of carbonates and silicates, CO_2 and SiO_2 remain to be dealt with.

The gangue material of iron ores usually consists mainly of silica, with varying proportion of alumina. There are usually present, also, various bases, such as lime, magnesia and oxides of manganese and the alkalis, which may or may not exist combined with the silica and alumina. These combinations of siliceous and basic substances may be in such proportions that they are self-fluxing, and hence require no added flux. Usually, however, there is a deficiency of bases, and at least part of the acid constituents,—silica, alumina, and sulphur,—must be supplied with them.

NOMENCLATURE OF IRON ORES.

The various ores of iron are designated by names which are more or less descriptive of prominent characteristics.

Magnetite.—Magnetite, Fe_3O_4, containing 72.4 per cent. Fe when pure, is so called because it is so strongly magnetic that it can be picked up by a magnet almost as readily as metallic iron itself. It is very heavy, having a specific gravity of 5.2, is generally very hard, is steel-gray in color, and gives a black streak when rubbed on

unglazed porcelain. It occurs frequently as octahedral crystals. Iron Age, Dec 17, 1903, p 10
One molecule is sometimes replaced by Mn., which would seem to
indicate that its constitution is $FeFe_2O_4$, a ferrous ferrate. It
has the same composition as the black scale which forms on iron
at temperatures above redness, as rolling-mill scale. Such scale
is the result of the action of the oxygen of the air upon the iron at
high temperatures.

Hematite.—Hematite, or Red Hematite, Fe_2O_3, containing
70 per cent of Fe when pure, has a deep red color and derives its
name from the Greek word " haima," meaning blood. It gives a
red streak to unglazed porcelain, and is only slightly affected by
a magnet. It occurs frequently in columnar formation, resembling
bundles of fibres, and also in a brilliant scaly structure, when it is
called specular hematite. Generally, however, it is granular or
massive, sometimes botryoidal or stalactitic, and often unctuous to
the touch It crystallizes in the rhombohedral system, and has
hardness and specific gravity about equal to that of magnetite. A
familiar form is the red scale which appears on cold finished
steel, or the deep red rust which attacks iron or steel in dry atmospheres It is identical in composition with common red iron-rust

Limonite.—Limonite, $2Fe_2O_33H_2O$ is a hydrous hematite,
containing 59.9 per cent. metallic iron and 14.5 per cent. H_2O of
combination. It gives a yellow streak on unglazed porcelain, and
hence is sometimes said to ·derive its name from the French
" limon," meaning lemon. As it is found largely-in bogs or mead-
ows, it is more probable that the name is derived from the Greek
" leimon," meaning meadow. It occurs usually in massive form,
showing frequent botryoidal, concretionary or mamillary forma-
tions. It is somewhat softer than hematite, and its specific gravity
falls below 4. It differs from red hematite only in the fact that it
contains water, and may therefore be considered a hydrous hematite. This water is not simply moisture that can be evaporated at
212 degrees F., but is water of combination which can be driven
off only at higher temperatures Its elimination destroys the
variety, however, and leaves the red hematite as the result of the
dehydration.
A considerable number of hydrous hematites, representing

various degrees of hydration, stand between hematite and limonite. They are of yellowish brown color, and give streaks varying
between the red of hematite and the yellow of limonite. They
are generally known as **Brown Hematites.** A familiar form of
limonite is the fresh yellow rust of newly-rusted iron, before it
has turned to the characteristic deep red color of maturity.

Siderite.—Siderite, $FeCO_3$, contains 48.3 per cent Fe and
41.4 per cent. CO_2. It is frequently known as Spathic Ore from
the resemblance of its cleavage to that of feldspar. It crystallizes
in the rhombohedral system, is softer than the oxides, of lower
specific gravity, and does not give a characteristic colored streak.
It changes to limonite and hematite on weathering. It is rarely
used as an ore in its raw state, but generally is subjected to a preliminary calcination to remove the CO_2. This leaves behind only
the iron and oxygen in the familiar form of hematite.

Silicates.—Silicates of Iron rarely occur in nature in quantity or richness sufficient to warrant their use as ores, but in
the form of cinder from heating and puddling furnaces and slags
from steel-melting or other refining furnaces, they frequently find
their way into the blast furnace. Such materials rarely carry more
than 55 per cent. of iron, and generally have 20 to 30 per cent.
SiO_2 to be fluxed. They are therefore productive of much slag
and may be advantageously mixed with rich ores which have a
deficiency of slag-making ingredients.

Other Elements in Iron Ores.—Other elements occur in iron
ores which have the most potent influence upon their value.
Chief of these are phosphorus and sulphur.

Phosphorus is usually present in ores of iron in the form of
a phosphate of calcium called apatite, which has the formula
$Ca_3P_2O_8$. Apatite frequently appears in the form of hexagonal
prisms, having pyramidal ends.. It has a vitreous to sub-resinous
lustre, and generally a pale brown color, but may be green, blue,
yellow, red or white. It frequently contains chlorine and fluorine.

Sulphur usually occurs in the form of pyrite or marcasite,
both sulphides of iron, having the formula FeS_2. They are
usually of a pale yellow color and a brassy lustre, and are gen-

erally crystallized, with striated faces The sulphur in the great Cornwall, Pa , deposit occurs as a cupriferous pyrite

Oxide of Manganese also occurs frequently as black layers or lumps scattered through ores of iron, particularly in the Southern States.

PHYSICAL CONDITION OF ORES.

The physical conditions of iron ores show quite as wide variation as the chemical compositions Not only does the product of almost every mine present distinct, characteristic features, but different parts of the same mine may yield ore of quite different quality and properties. The occurrence of ore ranges from soft beds of finely divided oxides that can be scooped up by steam shovels, to hard, flint-like deposits, that must be attacked with a diamond drill. As we shall see later, neither of these extremes is desirable The very hard deposits are expensive to mine, and the run of mine product is so coarse that it has to be crushed before charging in the furnace, which adds to the expense. When crushed, however, these ores are most desirable, as they give an open charge and small loss in flue-dust Fine ore, while easy to mine and handle, tends to cause irregularities in the operation of the furnace, and much of it is carried off by the gases. The magnetites of New York and New Jersey are of good physical condition, but they are so dense that it is said they cannot be used alone in the making of certain grades of iron without excessive consumption of fuel. The ideal ore is a lumpy ore that is not so dense but that it can be readily permeated by the furnace gases

VALUATION OF ORES

The value of an ore of iron depends upon many conditions Three conditions at least may be considered essential, the absence of any one of which may prove fatal to an ore otherwise desirable. These three requirements are: richness, accessible location, and suitable composition.

Richness. —By the richness of an ore is meant the percentage of metallic iron it contains The word " unit " is generally used to represent a per cent. of iron. An ore containing 60 per cent. of metallic iron is therefore said to contain 60 units. Since the metallic content of the ore is the only part that has a market value, it is clear that its proportion is of vital importance, since it must stand not only the costs of its own extraction, but also the additional fixed charges of mining and transportation. Other things being equal, however, the value of an ore does not advance at the same rate as the percentage of iron, but at a more rapid rate. This is because every additional pound of oxide of iron per ton of ore means one pound less of gangue to be fluxed, melted and handled. The price of ore increases therefore at arbitrary intervals. For example, if an ore ranging from 50 to 55 units is worth 6 cents per unit, one ranging from 55 to 60 units may be 7 cents, and 60 to 65 units, 8 cents per unit.

Location.—A rich ore may be valueless if its geographical location is remote or inaccessible. Under this head there are two fixed items to be considered in estimating the value of ores; namely, the cost of mining and the cost of transportation. A deposit of rich ore may be within the usual radius of transportation, yet its nature be such that the cost of mining is so high, compared with more tractable deposits, that it fails to prove a profitable undertaking. On the other hand, many rich and workable deposits in the West are so far from the centres of iron smelting that they will not be available until the more accessible supplies are exhausted, or until a demand is developed which permits of local consumption.

Composition. —It may happen that an ore, both rich and accessible, contains components which, on account of their injurious effects on pig iron, make it practically worthless. Since, however, not all of the non-ferrous components are harmful, it is necessary to know the nature and quantity of each before deciding the value of the ore.

The non-ferrous elements may for the moment be divided into three classes, according to their behavior toward the iron during smelting, namely, (1) those that never enter the iron, (2) those that partially enter the iron, (3) those that always enter the iron.

In the **first class** are the oxides of aluminum, calcium, magnesium and the alkalis. They may be dismissed as having no bearing on the ore value from this standpoint.

Under the **second class** come those elements which may enter the iron in quantities which vary as the furnace is operated, and which are therefore more or less under the control of the furnaceman, such as silicon, sulphur, and manganese.

The case of silicon is quickly disposed of. Since very little of the silicon in the iron is derived from the silica in the gangue materials, its presence does no harm beyond decreasing the richness of the ore.

The question of sulphur, however, is more serious. Ores which have been derived from the oxidation of pyritous deposits contain considerable quantities of residual sulphur. In thoroughly oxidized ores, the sulphur that may once have existed is diminished to a mere trace. A secondary impregnation of sulphides, however, may introduce a considerable quantity of sulphur into otherwise desirable ore. By proper furnace manipulation an ore running from $\frac{1}{2}$ to 1 per cent. sulphur may be used for making some grades of iron which are fairly low in sulphur. If sulphur exists in the ore in greater quantities, it is advisable to remove some of it by a preliminary operation of roasting. This operation, however, adds to the expense, and its cost must be known before the availability of the ore can be determined. The presence of even moderate quantities of sulphur demands additional flux and fuel for its removal, and hence its presence always lessens the value of ore.

Manganese, being somewhat more difficult to reduce from its oxides than iron, is never fully extracted from manganiferous ore. The quantity extracted is affected by the temperature and other conditions of the furnace hearth. Under ordinary conditions the extraction ranges from 50 to 75 per cent. of the manganese present. As we have seen, manganese in quantity ranging from 1 to 2 per cent. does not materially injure any grade of pig iron, and is a positive benefit to some grades. Therefore it is evident that manganese in an ore up to 2 per cent. of the iron present would not be a serious defect. If the percentage rose to 5 or 10, however, the ore could be used only as a fraction of a mixture with non-manganiferous ores. If the manganese rose to or above 15 or 20 per cent. of the iron present the ore would acquire a new value, as it would then be available for the production of such special products as spiegeleisen or ferromanganese and would no longer be considered as available for the production of pig iron

In the **third class** is the single element phosphorus. Of all the non-ferrous components of ore, phosphorus is most liable to affect its value. Its presence is the basis of the great division of iron ores into Bessemer and non-Bessemer ores It may be said briefly that this classification rests on the facts that, (1) practically all of the phosphorus present in the furnace enters the pig iron; (2) in the Bessemer process of making steel none of the phosphorus so accumulated can be eliminated; (3) Bessemer steel must contain less than 0.1 per cent. phosphorus. Therefore, since all of the phosphorus in the ore enters the iron and remains in the steel, it follows that an ore to be used for making iron for the Bessemer process must not contain more phosphorus than will make 0.1 per cent. of the resulting steel This amount approximates $1/_{1000}$ part of the iron in the ore.

The classification of ores as Bessemer and non-Bessemer was for many years the only distinction based on the content of phosphorus. With the development of other processes for making steel there have grown up other distinctions in pig iron, which have their origins in the phosphorus content of the ores.

We may distinguish five types of pig iron whose uses are determined primarily by their content of phosphorus.

(1) Phosphorus below 0.03 per cent., known as **low-phosphorus pig,** and used chiefly in making steel castings and low-phosphorus muck bar for crucible steel melting. It requires that the ores used should not average much over 0 01 per cent. phosphorus, and that only very low phosphorus coke and limestone should accompany them.

(2) Phosphorus below 0.1 per cent., or **Bessemer pig,** to be used in making rolled steel by the Bessemer or acid open hearth process. It requires that the ores used should not average over 0.05 per cent. phosphorus with average coke and limestone.

(3) Phosphorus below 0.2 per cent., **malleable pig,** to be used in making malleable iron castings. It requires that the ores used should not average over 0.1 per cent. phosphorus with average coke and limestone.

(4) Phosphorus below 1 per cent., **foundry** and **basic pig,** the

former for use in making iron castings, the latter for making steel by the basic open hearth process. They require that the ores used should not average over 0.5 per cent phosphorus with average coke and limestone.

(5) Two to 3 per cent phosphorus, known as **basic-Bessemer** or **Thomas-Gilchrist pig,** to be used for making steel by the basic-Bessemer or Thomas-Gilchrist process It requires that the ores should not average less than 1 to 1.5 per cent. phosphorus with average coke and limestone

RELATIVE VALUE OF ORES.

The relative values of ores of differing compositions depend upon the value of fuel and flux needed to smelt them Pig iron of a given composition always requires a constant quantity of metallic iron, whether that iron is extracted from a rich or a lean ore. A pig containing 95 per cent. metallic iron must always have 95 pounds of iron in each hundred of pig. If the ore chances to be lean, more of it must be used Disregarding the gangue for the moment, the cost of extracting 95 pounds of metallic iron evidently consists only of the value of the carbon needed to reduce and melt it. It is evident, then, that for a given fuel this cost must be constant for each ton of pig, no matter whether the ore be rich or poor. Therefore the relative values of ores are not affected by the cost of extracting their iron content, but center entirely in the relative expense of treating the gangues. The estimation of the cost of treating the gangue of an ore naturally divides itself into three steps:

(1) Determining the amount of flux needed to form the slag.

(2) Determining the amount of slag formed.

(3) Determining the amount of fuel needed to melt the slag.

It is necessary to have analyses of materials before such an estimate can be attempted. Let us assume, for example, an ore to have the following composition in the natural state:

	Per cent.			cent.
Fe	42.80	SiO		16 14
Mn	0.04	Al_2O_3		2 80
P	0.13	CaO		0 07
S	0.04	MgO		0 33
		Combined H_2O		12.84

It is desired to ascertain its value to the smelter when coke costs $3 per net ton and limestone $0.60 per gross ton at the furnace, and $2.30 represents all other expenses except the cost of the ore and $12 is to be the maximum cost of the pig iron. It is required, also, that the pig iron should contain 95 per cent. Fe and 1 per cent Si. Let us assume that it has been found by experience that results are attained best when the ratio of bases to acids in the slag is 1:1.2. Moreover, the limestone at hand contains 57 per cent of slag-forming materials, of which 50 per cent. is available base and represents its efficiency. The fuel affords 78 per cent. available carbon and requires 25 per cent. of limestone to flux its ash

[1] M. E., 1892.

As pointed out by Gordon, the heat developed in the average furnace hearth compared with the heat requirements shows that the reduction, impregnation and melting of pig iron usually requires 66 per cent. of carbon for its accomplishment. It is observed also that each pound of slag needs about a quarter of a pound of carbon to melt it.

In estimating the fuel needed by the above ore to produce a ton of pig iron with blast 1000 to 1200 degrees F., we can allow 0.66 tons of carbon for the iron in each ton of pig containing about 1 per cent. Si. For each extra pound of silicon 5 pounds additional carbon is necessary. In estimating the cost of fluxing and melting the gangue the first step is to determine the quantity of stone needed.

The balance sheet of the gangue is as follows

Acids		Bases	
SiO_2	16.14	MnO	0.02
Al_2O_3	2.80	CaO	0.08
		MgO	0.53
Total acids......	18.94		
SiO_2 reduced to Si......	0.91	Total bases.	0.63
Acids to be fluxed......	18.03		

$$\frac{1.2}{1} \text{ ratio } \frac{\text{bases}}{\text{acids}} \text{ in slag.}$$

36.06
18.03

Bases needed.................21.63
Bases present.................0.63

Bases to be supplied.............21.00

21.00 − 0.50 (efficiency of stone) = 42 per cent. of stone.

$\frac{95}{42.8} = 2.22$ gross tons of ore needed to make 1 ton of pig.

$2.22 \times 0.42 = 0.9325$ gross ton of stone needed to flux the gangue of 2.22 tons of the ore

Having found the quantity of stone needed, the next step is to find the weight of slag made by the ore   This amount will evidently be the sum of all the slag-forming ingredients in the gangue and the stone used to flux it, thus:

	Gross tons
Slag from the gangue (18 03 + 0 63) 2 22 =....	0.4142
Slag from the stone, 0.9325 × 0 57 =...	0 5315
	0 9457
The carbon needed to melt the slag, 0 9457 × 0 25 =	0 2364
The carbon needed to reduce, impregnate and melt the iron =..	0 6600
	0 8964

Of coke having 78 per cent available carbon there will be needed $\dfrac{0.8964}{0.78} = 1.15$ gross tons, or 2,575 pounds, per gross ton of pig.

The fuel needs 25 per cent. stone to flux its ash, hence the total stone needed will be found thus:

	Gross tons
Stone required by the gangue of the ore =..........	0 9325
Stone required by fuel, 1.15 × 0 25 =	0.2875
	1 2200

Allowing 5 per cent. loss of fuel as braize, 2575 pounds in the furnace represents 2700 pounds purchased at $3 per net ton, hence:

The fuel per ton of pig iron will cost..................	$4.05
1.22 gross tons stone, at 60 cents per ton .	0 73
The fixed charge, salaries, wages, supplies, repairs, &c.	2 30
Total cost of pig iron, exclusive of ore...	$7.08
Total cost of pig iron allowable..........	$12 00
Total cost of pig iron, exclusive of ore.	7 08
Value of 2 22 gross tons of ore.	$4.92

or $2.20 per gross ton, delivered

Such an ore could be valued at 5 cents per unit on a basis of 43 per cent. Fe.   A premium of 7½ cents per unit above 43 might be paid and a penalty of 10 cents for each unit below imposed, with the understanding that no ore below 40 per cent. would be acceptable.

For other methods of estimation of all raw material used in the blast furnace see *The Iron Age,* April 14, 1904, p. 12.

PREPARATION OF ORES.

By far the greater proportion of ores come from the mines in such condition that they may be charged into the furnace at once Certain kinds, however, must be put through some preliminary process in order to render them more tractable in the furnace. The most usual preparatory processes are calcination, roasting, concentration and agglomeration

CALCINATION.

Calcination of ores has two objects, namely.

> (1) To prepare them for better reduction by changing their physical properties, either
>> a) by removing water from hydrous ores;
>> b) by removing CO_2 from spathic ores,
>> c) by making dense ores more permeable to gases.
> (2) To render the ore magnetic, in order to facilitate subsequent concentration.

(1a) Usually the **removal of water** alone by calcination is superfluous, as the water is readily evaporated in the top of the furnace by heat that would otherwise be wasted.

(1b) The **removal of carbon dioxide** before charging is more imperative. According to Wedding, carbon dioxide can be completely eliminated from ferrous carbonate by heat, either with or without access of air. The reactions are slightly different, however. With access of air, $FeCO_3$ decomposes as follows:

$$6FeCO_3 + 3O = 3Fe_2O_3 + 6CO_2.$$

Without air, the reaction runs thus:

$$6FeCO_3 = Fe_6O_7 + 5CO_2 + CO.$$

The expulsion of CO_2 from $FeCO_3$ absorbs 465 B. T. U. per pound of carbonate. Hence, it should be done outside of the furnace, where the necessary heat can be produced more cheaply.

(1c) Calcining **oxide ores** is beneficial only to the physical condition of the ores. It may facilitate the reduction of dense ores by making them more accessible to the furnace gases This is especially true of magnetites having dark gangue material, as that usually indicates a ferrous silicate, which is capable of oxidation. Ores which can be broken only with difficulty before calcination

Inst Jour;
1806, II.,
p 116

are sometimes disintegrated completely by the operation. The degree of oxidation of the ore will depend upon the temperature of the treatment. Heating Fe_2O_3 to high temperatures in air will produce Fe_3O_4 or Fe_6O_7, which are strongly magnetic. Continued heating in air for sufficient period at low redness will restore the original Fe_2O_3 and the magnetic property will be lost. The reduction of Fe_2O_3 in the furnace is more easy than that of Fe_3O_4 and Fe_6O_7. Yet for certain reasons these compounds may be necessary. If it is desirable to concentrate the ore by means of magnets, these magnetic oxides are essential, since Fe_2O_3 is only slightly susceptible to magnetism, and FeO cannot be produced commercially.

At Wharton, N. J., the Hibernia ore, a dense magnetite, is calcined with gas in a Davis-Colby kiln for the purpose of changing it to Fe_2O_4 and thereby rendering it more susceptible to furnace reactions.

(2) Calcination for the purpose of **rendering ores magnetic** is unnecessary in the case of magnetites, as they are already magnetic. Spathic ores, as we have seen, when decomposed in a neutral atmosphere, yield the magnetic oxide, Fe_6O_7. If they are decomposed at sufficiently high temperatures in air, they will yield Fe_3O_4.

Hematites may be rendered magnetic in two ways.

a) By heating strongly in the absence of air, oxygen is liberated thus:

$$3Fe_2O_3 = 2Fe_3O_4 + O = Fe_6O_7 + 2O.$$

The temperature must be very high, and it is difficult to prevent sintering of the more fusible components.

b) By heating in the presence of reducing agents, some of the oxygen may be removed at moderate temperatures, leaving the residue magnetic. This may be accomplished by the use of solid carbonaceous matter, or by means of reducing gases, such as carbonic oxide or hydrogen, or by means of water or producer-gases, which contain them. Limonites and brown hematites, when dehydrated, act very much as hematites. They are more easily attacked, owing to their porosity.

The calcination of lump ore is best performed in vertical

shaft furnaces with a good batter to facilitate the descent of the ore. The air for the combustion of the fuel should be so regulated that it consumes only enough of the fuel to keep up the necessary temperature, thereby leaving a residue to act as a reducing agent. The desirable temperature is dependent upon the composition of the ore. If the gangue is fusible, moderate temperatures must be used.

Since the magnetization of a piece of ore must proceed from the outside inward, the lumps must not be too large, or the action will be incomplete, or unnecessary time will be consumed. On the other hand, fine ores are difficult to magnetize uniformly, as the gases make channels in the ore, through which they pass, leaving untouched material on either side. Finally, it is desirable that the material should be approximately uniform in size, in order that the fine material may not clog the spaces between the coarse, and thereby impede the free passage of the gases.

ROASTING.

The object of roasting is the removal of sulphur. It consists essentially of heating the ore to a high temperature with contact of air. The sulphur is present usually as pyrite, which has the formula FeS_2, showing that two atoms of sulphur are in combination with one of iron. One of the atoms of sulphur is loosely attached and can be expelled by simply heating without contact of air, thus:

$$FeS_2 = FeS + S.$$

The elemental sulphur thus liberated escapes by volatilization. Heat alone is insufficient to decompose the FeS which remains. By the addition of oxygen from the air, however, a further change takes place, and ferrous sulphate is formed, thus:

$$FeS + 2O_2 = FeSO_4.$$

Further heating completely decomposes the sulphate, forming oxide with the liberation of sulphur dioxide and oxygen, thus:

$$2FeSO_4 = Fe_2O_3 + 2SO_2 + O.$$

These two conditions are well illustrated by experiments by Valentine.

Substance.	Atmosphere	Temperature degrees F	Time, hours,	S lost. Per cent	S as sulphide in residue Per cent	
Pyrite....	With air.	1,200	4	98	49	Tr A I. XVIII.,
Pyrite ..	3CO+1CO₂	1,800	4	37	99	
Pyrite .	Without air.	2,500	18	44	99	

From these experiments the following conclusions are evident:

1. Sulphur is almost completely removed from FeS_2 by heating in air at moderate temperatures.

2 Prolonged heating of FeS_2 at high temperatures out of contact with air or in an atmosphere of furnace gases will not remove over 50 per cent of the sulphur.

3. The sulphur in the slight residues of (1) exists largely in the condition of sulphate; the sulphur in the heavy residues of (2) and (3) exists as practically unchanged sulphide.

Although these experiments were made on pure pyrite, they serve well to illustrate the general effects of roasting.

Experiments in roasting Cornwall ore, containing 2.66 per cent S in the presence of air under varying conditions gave these results.

Ore	Per cent. S.	Temperature, degrees F.	Time, hours.	S lost Per cent.	S as sulphide in residue. Per cent	
Cornwall...............	2 66	1,200	2	87	32	Ibid
Cornwall... .	2 66	1,200	4	93	51	
Cornwall...............	2 66	1,500	1	96	72	
Cornwall.......	2.66	2,400	¾	9	3½	

From these figures it appears that the higher the temperature and the longer the exposure at a given temperature, the more complete the elimination, within certain limits. If the temperature is raised high very suddenly, as in the fourth experiment, the sulphide fuses, thereby retarding oxidation and preventing volatilization. The prime requisites for successful roasting are control of temperature and excess of air.

Operation.—In roasting ores on a large scale as a preparation for smelting, it is desirable to have a temperature sufficient for successful oxidation with the least expenditure of fuel. Solid, liquid or gaseous fuel may be used. The solid fuel is mixed with the charge of ore in a suitable kiln and caused to burn in contact with the ore. In this way the ore may be roasted in tolerably thick bodies. Liquid and gaseous fuels are burned in a com-

bustion chamber at one side of a comparatively thin body of ore.
The products of combustion pass through the column on the way
to the chimney.

An example of the first kind of roaster is the **Gjers kiln,** or
Cleveland Calciner, as it is used in the Cleveland iron district of
England, and at Lebanon, Pa. It consists of a cylindrical steel
plate jacket contracted toward the bottom, where it is provided
with outlets for the product and inlets for air. The whole is then
lined with fire bricks. According to Turner, such a kiln, 33 feet
high and with 24 feet diameter, would have a weekly capacity of
nearly 1000 tons. The ore, mixed with about 5 per cent. of small
coal, is charged at intervals at the top and withdrawn at the
bottom. The rate of progress depends upon the rate of withdrawal, but it should not exceed about 12 feet per day, which is
equivalent to a daily product of about 140 tons. As used at
Cornwall, Pa., the roasters ranged from 10 feet to 20 feet high,
and 12 to 22 feet in diameter, and used 50 to 100 pounds coal
per ton of ore, roasting 10 to 50 tons per day, according to the
size of kiln. The time of treatment was from two to ten days,
and the cost ranged from 20 to 70 cents per ton. The average
was 30 cents, 12 cents for fuel and 18 for labor.

An example of the second type of roaster is the **Davis-Colby
kiln,** which is largely used for the ores of Cornwall, Pa. The
earlier form of this kiln consisted of a central cylindrical flue,
which led to the chimney, and was surrounded by an annular
space, about 18 inches wide at the top, and increasing slightly
downwards. Firebrick walls enclosed the annular space, and
separated it from the central flue. The ore was charged in the top
of the annular space, and gas and air were admitted through the
surrounding wall. The gas burned in a combustion chamber, and
the products of combustion were drawn through the ore into the
central flue by the draught of the chimney. The roasted ore was
withdrawn at the bottom

A comparison of results produced by this method with those
of the Gjers kiln when both were working on the Cornwall ore
showed a decided advantage in favor of the gas method The
sulphur is more nearly eliminated, and what remains is more fully
changed to sulphate, thus:

Kiln.	Per cent S in product.	Per cent S as sulphate.	Per cent S as sulphide.	
Gjers	{ 1.133 { 1 380	24.98 9.78	75 02 90 22	
Average	1,256	17.38	82.62	Ibid.
Davis-Colby	{ 0.782 { 0 798 { 0 596	54.21 39.85 49.14	45.79 60.15 50 86	
Average	.0 725	47.73	52.27	

More recent experiments show results as low as 0.3 to 0.4 per cent. S., with an average of 0.8 to 0.9 per cent.

The later forms of the Davis-Colby kilns are rectangular in shape and of large dimensions. The kiln at Wharton, N. J., is 100 feet long by 36 feet wide, having receiving bins on top and discharge chutes beneath. Three lines of standard gauge tracks run along the top of the bin, 55 feet from the ground The body of the kiln is raised from the ground 19 feet, to permit of the construction of chutes, which discharge the roasted ore into railroad cars. It is built in duplicate and consists of two rectangular chambers the length of the kiln, 24 feet deep and about 12 feet wide. These chambers are divided longitudinally by two brick walls, making three spaces which connect at intervals. The middle one, which has a slight batter downward, is the ore chamber. The outer space serves as a combustion chamber and is fed from a supply pipe by numerous gas burners at intervals near the bottom. The inner space serves as a collecting chamber for the products of combustion which have passed through the ore. The two inner chambers connect with a central flue that leads to the chimney. This kiln is used to calcine dense magnetic ores, to render them porous. It is said to treat 600 to 800 tons per day at a temperature of about 1200 degrees F. Similar kilns are in use at Lebanon also, for roasting sulphurous ore.

CONCENTRATION.

The concentration of ore has two objects, which may or may not be simultaneous, namely, enrichment, and the removal of objectionable substances. Concentration may be of two kinds, wet or dry.

Wet concentration is usually applied to ores which contain

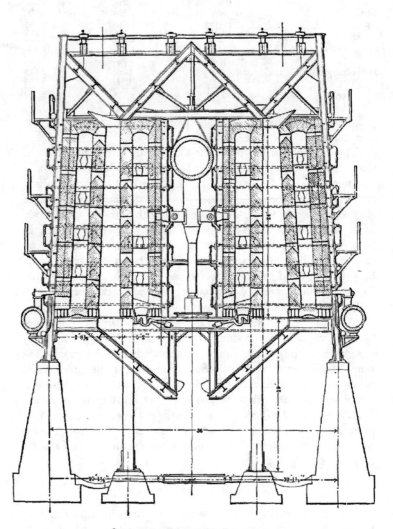

Davis-Colby Kiln at Wharton, N. J.

clay, pebbles or sand. It may be of two kinds, namely, washing and jigging. The **washing** process is applicable to ores in which clay is the objectionable constituent. The usual form of washery consists of a log, carrying numerous fins or scrapers on its surface, and revolving in an inclined trough which is plentifully supplied with water. The ore which is charged into the lower end of the trough is constantly worked by the revolving fins toward the head of the trough, where it is discharged, and the clay, after being thoroughly disintegrated by the fins, is carried off by the water at the foot of the trough.

Tr. A. I. M. E., XXIV, p 84

Ores which contain pebbles or sand must be treated by the **jigging** process, since the particles are too heavy to be carried away by gently flowing water. The jigs consist of boxes with perforated bottoms, set in tanks of water. The water, pulsating rapidly through the perforations, causes the different particles in the boxes to separate into layers according to their respective specific gravities. The ore collects on the bottom, that which is fine enough passes through the perforations into the tank, while the pebbles and sand which lie on top are allowed to overflow and go to waste. Both washing and jigging find their chief application in the local ores of Pennsylvania, Virginia and Alabama. Washing is much more widely practised than jigging, but even washing is applied to only a very small percentage of the ores used.

Tr A. I. M. E, XIII, p. 35.

Dry Concentration.—Dry concentration usually takes the form of magnetic separation. The application of this process depends upon the fact that all ores are magnetic, or may be made so. When the ore is crushed to such a degree of fineness that the compounds of iron are detached from the gangue material and the mixture is passed in a thin layer before strong magnets, the magnetic particles are attracted away from the non-magnetic, and a more or less complete separation takes place. If the ore is not magnetic it is necessary to make it so before applying this process. Some ores may be magnetized simply by heating at certain temperatures, but generally it is necessary to heat in the presence of finely divided carbon, such as coal, or other reducing substances. Partial reduction of completely oxidized ores tends to make them more magnetic.

There are several types of magnetic separators, which differ more in construction than in principle.

The **Ball-Norton** " drum type " separator consists of horizontal drums revolving around stationary magnets which act through the lower third of the drum. The ore is fed beneath the drums, which hold the magnetic material, while the tailings fall to the ground. As the drum rotates, the magnetic material clinging to it is carried out of the field of the stationary magnet and the load falls into a bin, or into the hopper of the next drum.

Tr A. I. M. E.,
XIX, p. 187.

The **Wenstrom** separator also has revolving horizontal drums, but the magnets are placed at the side instead of the bottom The ore drops to the magnetized area, whereupon the tailings fall to the ground, while the magnetic particles are attracted to the drum and carried to a bin beneath it.

Ibid,
p. 65.

The **Buchanan** separator consists of two rolls revolving toward each other and connected by a horseshoe magnet. The ore is fed between. The non-magnetic tailings fall to the ground, while the magnetic particles cling to the rolls and are conveyed to bins beneath.

Ibid,

The **Ball-Norton** " belt-type " magnetic separator is in use at Mineville, New York, where it is said to surpass the drum type in capacity and efficiency, beside being cheaper to construct and maintain. It consists of a series of twelve magnets of alternate polarity, placed side by side above a moving belt, which is well within the magnetic zone. A feed belt running on a slightly lower level brings the material under the magnetized belt, as far as the third magnet The magnetic particles adhere to the under side of the magnetized belt, while the non-magnetic material is carried on to the tailings bin. The magnetic portion is conveyed to its bin across the faces of the magnets, and is turned end over end by their alternating polarity. This action tends to free entrained non-magnetic particles, thereby making a purer product. Each machine is said to be able to work 30 tons per hour.

Iron Age,
Nov 23, 1905,
p 1367

The **Wetherill** separator is made in two distinct types, both of which are used at Mineville.

The " roller " type, which is designed for the selection of strongly magnetic material, consists essentially of a horizontal roller revolving between the poles of a stationary electro-magnet. The roller becomes strongly magnetic through induction. The crushed ore is fed on the top of the roller and is carried by the revolution into the line of magnetism. The non-magnetic minerals fall from the roller into the tailings chute, and the magnetic portion is carried forward to the neutral point, where it drops into the concentration chute. Machines are built to operate on 6, 14 and 30 amperes at 110 volts, and are said to handle 400 tons per 24 hours.

The " cross-belt " type, designed for selecting weakly magnetic

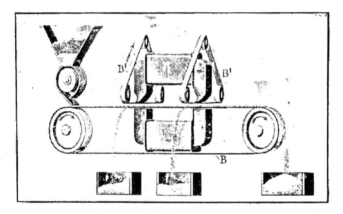

The Wetherill Magnetic Separator, Type E.

minerals, consists of two horseshoe electro-magnets, placed horizontally with poles facing, the pole pieces on the upper magnet being wedge-shaped, and those on the lower, flat. Between the magnets runs a horizontal belt which carries the material. Beneath each of the wedge-shaped poles and just above the material runs a small cross-belt. The paramagnetic minerals jump toward the wedge-shaped pole where the lines of force are concentrated. They adhere to the underside of the cross-belts and are carried out of the magnetic field, where they drop by gravity. The diamagnetic materials are carried on by the feed belt to the tailings bin. This type is made in capacities ranging from 6 to 30 amperes at 110 volts pressure.

Magnetic concentration of iron ores has been most fully developed in this country at Mineville, where the operation is carried out on two grades of ore; in one case for the purpose of enrichment, in the other in order to separate an undesirable constituent. The two grades are: (1) The Old Bed ore, which consists of magnetite already rich in iron, but containing a prohibitive amount of phosphorus, and (2) the New Bed and Harmony ores, which are a low grade magnetite with a moderate amount of phosphorus The results of concentration of Old Bed ores are a very rich concentrate, reasonably low in phosphorus, and tailings low in iron, but high enough in phosphorus to have a value as fertilizers. The results of concentrating the New Bed and Harmony ores are a rich concentrate with still lower phosphorus and valueless tailings. Fortunately, the ores are of granular structure and the phosphorus exists in the form of apatite, or calcic phosphate, which is more or less completely freed from the oxide of iron by crushing The method of procedure is to crush the ore so that it will all pass a six mesh screen. The crushing is done by means of a Blake rock crusher and a set of Reliance rolls. The crushed ore is then separated by screening into portions that will pass through the 30, 16, 10 and 6-mesh screens respectively. These are then dried in an Edison drying tower and sent to Rowand and Ball-Norton magnetic separators. The tailings from this operation are sized again to 16 and 20 mesh, and re-treated on the Wenstrom and Wetherill magnetic separators The oversize is crushed again in a set of Reliance rolls, and re-treated. The plants for treating the Old Bed and Harmony ores are practically duplicates, and the results of each are summarized below.

ᴵ Age, 1903, p. 16

Product	Per cent. Fe.	Per cent. P.	Per cent. ore.
Old Bed ore	59.59	1 74	100 0
Iron concentrates	67.34	0 675	83.0
No. 1 phosphate concentrates.	3 55	12 71	7 5
No 2 phosphate concentrates	12.14	8.06	7.5
Harmony ore	50 26	0 295	100 0
Iron concentrates	64.10	0 133	77.0
Tailings	13 97	0 877	23.0

A somewhat different method of procedure is in use at Hibernia, New Jersey. The ore is passed first through a Buchanan crusher and set of rolls, and crushed to a size which will pass a

2½-inch screen It is then passed over a Ball magnetic separator, having the low potential of 13 volts, which selects the richest of the ore, giving a coarse concentrate. The tailings from this operation are passed more slowly over a more powerful magnet of 25 volts potential, which selects all that is worthy of further treatment. The tails from this stage are allowed to go to waste, while the heads are recrushed and passed over a third separator having a potential of 15 This operation yields the fine concentrates

This differential method of separation saves much expense of crushing, and leaves the concentrates in a coarse condition, suitable for charging without agglomerating. The operation as carried out at Hibernia, on an ore containing only 47 per cent. iron, yields concentrates containing 60 per cent iron, and tailings that carry only 15 per cent.

AGGLOMERATION.

A very grave objection to fine ore is its tendency to be carried out of the furnace top by the escaping gases. This feature becomes more marked with high pressures of the blast and rapid driving of the furnace. For this reason fine ore is sometimes put through the process of agglomeration. This consists of assembling the fine particles into balls or cakes, which are held together by some adhesive.

Briquetting.—When admixed adhesives, such as fused slag, milk of lime, coal tar, rosin, molasses, malt liquors, glutens and various commercial preparations are used as-binders, the process is called " briquetting." A suitable binder must make briquettes that can be handled roughly without breaking or crumbling, that can stand changes in weather, and that will not be detrimental to the furnace or its product. The method of manufacture consists of mixing the required proportion of the binder thoroughly with the ore in a pugmill or other suitable device, of passing the mixture into a moulding machine, such as a brick press, which gives it shape, and finally of drying the briquettes in a suitable oven. Up to the present, however, briquetting has not proved a commercial success.

Nodulizing.—The process of nodulizing fine ores as developed by the National Metallurgical Company, of New York, con-

Iron Age,
Sept. 7, 1905
p.589.
sists in passing fine ores, flue dust, etc., through a rotary kiln at
high temperatures. The kiln consists of a shell of steel plates, riveted together and lined with bricks. It is set at a slight inclination
from the horizontal, and is 100 feet long by 6 feet diameter at
the feed end and 7 feet at the discharge end. The ore, mixed
with 1 per cent. of tar, is delivered by an automatic feeder into
the upper end, and finely powdered and dried coal is blown in by
a blast at the lower end. The ore is 1½ hours passing through
the kiln, and is discharged in the form of sintered and purified
nodules. The size of the nodules may be varied by varying the
temperature, the quantity of tar and the speed of revolution.
The tar serves as a binder until nodulizing begins, when it volatilizes, taking with it much of the sulphur and arsenic, and leaving
the product very nearly pure. The plant treats 175 tons in 24
Inst Jour.,
1906, III.,
p. 358.
hours, at an expenditure of 150 horsepower. The product is in a
granular to slightly lumpy condition, admirably adapted to use
in the furnace.

ORES OF THE UNITED STATES.

The ores of the United States which have commercial value
are of three kinds, hematites, limonites and magnetites.

In the northern states, the ores used are chiefly hematites from
the rich deposits of the Lake Superior districts. These ores are
red hematites, though sometimes hydrous. They average 54 per
cent. metallic iron, with little gangue, and require no preparation.
This enables them to compete with poorer local ores in all the
northern furnace districts. In some localities, however, the native
supply of brown hematites still furnishes a portion of the ores
used. In the eastern district, notably New York and New Jersey,
a considerable quantity of magnetites is mined for local use.
These ores are sometimes used as they come from the mine, and
sometimes are subjected to preparatory operations.

In the southern states, especially in Alabama, the ore supply
is of two kinds, hematite and limonite. The hematites are of
Clinton age, and above water-level exist as soft and tolerably rich
ores. In depth they are leaner, and are associated with 12 to
20 per cent. limestone, which renders them almost self-fluxing.
The limonite ores are of better quality, although they must be

subjected to washing before use. They average nearly 50 per cent metallic iron

Hematite Production.—During 1907, 46,060,486 tons of hematite ore, amounting to 89 per cent. of the total production, was mined in the United States Of this amount, 56 per cent. came from Minnesota, 23 per cent. from Michigan and the remainder from Alabama, Wisconsin and Tennessee. Over 90 per cent. of all the hematite produced, viz., 41,604,454 tons, was produced by what is commonly known as the "Lake Superior district." This district lies to the southwest and north of Lake Superior, and includes the northern portions of Michigan, Wisconsin and Minnesota, and part of Canada. There are five distinct groups of mines or ranges. Their relative shipments in 1906, 1907 and 1908 were as follows:

	1906. Long tons	1907 Long tons	1908 Long tons.
Mesaba	23,819,029	27,495,708	17,257,350
Menominee	5,109,088	4,964,728	2,679,156
Marquette	4,057,187	4,388,073	2,414,632
Gogebic	3,643,514	3,637,102	2,699,856
Vermilion	1,792,355	1,685,287	841,544
Totals	38,421,173	42,170,878	25,892,538

Of these five ranges, the Mesaba, although by far the most productive, is of comparatively recent development, and of radically different character from the others. The other four are generally referred to as the "old ranges" to distinguish them from the Mesaba.

The first range to be developed was the **Marquette.** It is situated in the upper peninsula of Michigan, with its eastern extremity on the lake shore at Marquette, which serves as its shipping-point. It extends westward for a distance of about 30 miles. It is most actively developed in the vicinity of the towns of Negaunee, Ishpeming and Champion. There is considerable production also at Republic, a few miles south of Champion. This region was discovered in the forties, and became an important producer in 1854 The ores are both hard and soft. The hard ores are massive or specular hematites, and must be drilled and blasted, thus forming lump ore. The soft ores are somewhat hydrated, and resemble limonite. Both Bessemer and non-Bessemer ores are produced.

The **Menominee Range** was opened in 1877. It is located in the vicinity of the Menominee River, in both Michigan and Wisconsin. The most active operation is at Iron Mountain, Michigan, which is about 40 miles south of the Marquette group, and about the same distance from the ports, Escanaba and Marquette, which serve as shipping points. The ores are generally soft, blue hematites, showing specular particles

The **Gogebic Range,** which was opened in 1885. is situated also on the boundary between Upper Michigan and Wisconsin about 10 miles from Lake Superior and about 100 miles from Marquette. It is about 20 miles in length, and is most actively worked at Bessemer and Ironwood, in Michigan, and at Hurley, Wisconsin The ores of these deposits are soft blue, brown, and black hematites, often high in manganese, and practically all Bessemer.

The **Vermilion Range,** which was also discovered in 1885, is quite remote from the three preceding ranges. It is located in northeastern Minnesota, about 75 miles due north from Duluth. It has two points of development, namely, at Soudan, near Tower, and at Ely, about 20 miles northeast of Tower. The ore found at Soudan is hard, dense, somewhat specular hematite. It is exceedingly difficult to drill Part is Bessemer and part is non-Bessemer. At Ely the ores are all soft and of good Bessemer quality.

The **Mesaba Range,** which is now such a heavy producer of iron ore, did not become a factor until 1893 It differs materially from the older ranges. Its location is about 20 miles southwest from Tower, and about 60 miles from Duluth. It is about 40 miles long, and has important developments at Biwabik, Virginia, Mountain Iron and Hibbing The ores are red, brown and yellow, of soft, loamy texture, and sometimes of extreme fineness. They are easily carried out of the furnace by the gases, and they also cause a deposition of carbon which obstructs the passage of the gases. Some of the ores are of Bessemer quality.

The **Clinton or Fossil Ores** derive their names from Clinton, N. Y., where they were formerly of considerable economic importance, and from the fact that they contain many fossil remains of the Upper Silurian Age They occur interstratified with shales and limestone They are scattered over a very wide area and are remarkably persistent. Wherever the Clinton stage appears, one or more beds of ore is present. The Clinton ore has reached its greatest economic development in the Birmingham district in Alabama, where folds and faults have brought the ore beds into close proximity to the coal and limestone of the region, thus making the working very economical Besides in Alabama, they have economic importance in East Tennessee, Northwest Georgia, Southwest Virginia, Maryland, Central Pennsylvania, and New York. They appear also in Kentucky, Ohio, and Wisconsin. The structure of the ore varies in the different localities, sometimes appearing as a replacement of fossils, again as oolitic concretions, and again as ferruginous limestone. No less than six separate beds appear at times. The richness of the deposits varies widely. They are always too high in phosphorus to be classed as Bessemer ores.

Brown Hematites and Limonites.—Next to the Red Hematites, although far below them in the amount produced, stand the Brown Hematites. In 1907, 2,957,477 tons of Brown Hematites were mined The greater part of this quantity was produced from the Siluro-Cambrian deposits of Alabama, Virginia and Georgia. These deposits occur in the slates, schists and limestones of the Cambrian and lower Silurian systems of the Appalachians. They are remarkably persistent and extend from Vermont to Alabama. They appear with considerable economic importance in eastern New York, Pennsylvania, central Maryland, southwestern Virginia, eastern Tennessee, northwestern Georgia and central Alabama. This belt formerly had very active development in the northern states, especially in Pennsylvania. It is still worked there to some extent, and also in Virginia and Tennessee, but its greatest development is now in Alabama, where it furnishes a large proportion of the ores used.

Limonite occurs in a number of localities in Colorado. The mines in Saguache County supply a part of the ore used at the Pueblo works.

Bog Ore is the name for the beds of limonite which form in pools of stagnant water Ferruginous waters coming in contact with alkaline or carbonated waters, precipitate their burden of iron from solution, and in time considerable beds of ore form. Such deposits exist in North Carolina, Canada and the West, but are not of present economic importance.

Magnetite.—Next in order of importance comes magnetite. In 1907, 2,679,067 tons of magnetic ore and of magnetic concentrates were used in the East. They came mostly from eastern New York, northern New Jersey, and Pennsylvania, where they occur in a series of lenses, roughly parallel in metamorphosed crystalline rocks, particularly gneisses. The ores are usually dense and hard, but show generally a granular, semi-crystalline condition which is favorable to concentration by magnetization. They are generally non-Bessemer, but in some instances may be brought to Bessemer quality by concentration, in which a portion of the apatite, which is the phosphorus-bearing mineral, is left behind. In the gneisses of the western counties of Virginia and North Carolina are found magnetite beds of Bessemer quality.

A unique and remarkable deposit of magnetic ore occurs at **Cornwall, Pa.** It is quite distinct from the gneissic deposits, being associated with green pyritous shales. The ore is a soft, earthy magnetite of low grade, but owing to the interlaminated condition of the shales, it does not lend itself readily to magnetic separation. It is very low in phosphorus and is used in Bessemer mixtures. Owing to the presence of pyrite and chalco-pyrite, however, the raw ore contains 2 to 3 per cent. sulphur, and is, therefore, roasted before smelting. The presence of the chalco-pyrite introduces nearly 1 per cent. of copper into all pig iron which is made entirely from these ores. This deposit has been worked continuously for more than a century, and has yielded many million tons of ore. It is apparently far from exhausted, and now reaches an annual output of three-quarters of a million tons.

In Colorado several magnetic deposits are known, but those in Chaffee County are the chief producers. They furnish a part of the ore smelted at Pueblo. Magnetic deposits occur also in Utah, California, and to some extent in the Lake Superior districts.

Immense quantities of iron are known to exist in the form of titaniferous magnetites, notably in New York, New Jersey and

Canada, and also in North Carolina, Minnesota, Wyoming and California, but this supply has not been available because of the difficulties hitherto ascribed to the melting of titaniferous slags. These deposits are usually low in phosphorus and sulphur, and when smelted are said to yield iron of remarkable excellence.

Siderite.—The production of spathic iron ore in the United States is of trifling importance, amounting to only 23,589 tons in 1907. This form of ore is the least desirable in the furnace, owing to its high content of carbon dioxide and consequent low content of iron. It often occurs associated with other carbonates, especially calcic and magnesic, which makes the ore at least partly self-fluxing. It usually appears as concretions embedded in strata of the carboniferous age. When the association is with bituminous matter, such as coal seams, the ore is known as "black band;" when embedded in shales and associated with much clay it goes under the name of "clay ironstone." The former was once of considerable importance in Pennsylvania, West Virginia and Colorado, and the latter in New York and Connecticut.

According to the annual report for 1908 of the United States Geological Survey, the following is the output in gross tons of iron ore from the chief producing States and a comparison of the outputs of 1907 and 1908:

State.	1907	1908	Kinds of ore in 1907
Minnesota	28,969,658	18,632,220	All red hematite.
Michigan	11,830,342	8,839,199	All red hematite
Alabama	4,039,453	3,734,438	77 8 per cent. red hematite. 22 2 per cent. brown hematite.
New York	1,375,020	697,473	89 per cent. magnetite
Virginia	786,856	692,223	88 per cent. brown ore.
Tennessee	813,690	635,343	67 per cent brown ore, 33 per cent. red hematite
New Jersey	549,760	394,767	91 per cent magnetite
Wisconsin	838,744	733,993	96 per cent. red hematite
Pennsylvania	837,287	443,161	84 per cent magnetite, 13 per cent. brown ore
Montana, Nevada, New Mexico, Utah and Wyoming	819,544	528,625	74 per cent red hematite, 24 per cent. magnetite
Georgia	444,114	321,060	76 per cent. brown ore, 24 per cent. red hematite.
Arkansas and Texas	118,667	55,966	All brown ore.
Missouri	111,768	98,414	Brown and red hematite

	1905	1906	1907	1908
Total product	42,326,133	47,749,728	51,720,619	35,983,336
Imported	845,651	1,060,390	1,229,168	776,898
Exported	208,017	265,240	278,208	309,099
Apparent consumption	43,433,138	49,355,343	51,880,398	36,451,135

68 *Blast Furnace.*

The following comparison by the same authority of the production and consumption of ore for the past ten years is instructive in showing the rapid increase of ferrous products in this country:

Year.	Ore mined Gross tons.	Pig made Gross tons.	Year	Ore mined Gross tons.	Pig made. Gross tons.
1895	15,957,614	9,446,307	1902	35,554,135	17,821,307
1896	16,005,449	8,623,127	1903	35,019,308	18,009,252
1897	17,518,046	9,652,680	1904	27,644,330	16,497,033
1898	19,433,716	11,733,934	1905	42,526,133	22,992,380
1899	24,683,173	13,620,703	1906	47,749,728	25,307,191
1900	27,553,161	13,789,242	1907	51,720,619	25,781,361
1901	28,887,479	15,878,354	1908	35,983,336	15,936,018

Foreign Ores.—In addition to the native ores of the United States, some imported ore is used in the furnaces along the Atlantic seaboard whose distance from ports is not so great as to preclude competition with Lake Superior ores. The chief sources of imported ores are Cuba, Spain and Newfoundland.

ANALYSIS OF IRON ORES.

Hematites

Mine and locality.	Fe.	SiO₂	Al₂O₃	CaO.	MgO	Mn.	P.	S	Loss by igni-tion	Moist-ure
*Angeline, Marquette.	57 94	3 07	1 17	0 13	0 08	0 27	0.040	0.011	1.80	10.78
*Cleveland Cliffs, Marquette	63 75	3 04	1 95	0 70	0.70	0.34	0.102	0 018	0 65	0.38
'Champion, Marquette	63.44	4 51	2.36	0 32	0 29	0 20	0 060	0.013	...	0.88
*Republic, Marquette	67 30	1 80	0.42	0.49	0 13	0.55	0.052	0 053	.0 00	0 59
*Chapin, Menominee	54.08	5.76	1 31	1.15	3.31	0.48	0.060	0.017	3 00	6 96
*Pewabic, Menominee	58 35	4.53	0.92	0.37	1 11	0.13	0.008	0 003	0.84	8 58
*Toledo, Menominee	45.20	24.19	1.05	0.49	1.13	0.10	0.006	0.010	0 95	7.16
*Ashland, Gogebic	53 49	6 34	2.70	0 38	0 24	0.30	0 040	0 009	2.54	10 70
*Cary Empire, Gogebic	52.47	5 56	0.79	0.17	0 26	2.26	0.053	0 005	4 77	10 16
*Chandler, Vermilion	60.92	3.97	2.08	0 60	0 13	0 21	0.038	0.002	0 85	5 54
*Vermilion, Vermilion	65 39	2.56	0.84	0 62	0.30	0 04	0.086	Tr.	0.30	1 56
*Biwabik, Mesaba	56.93	3 67	1.21	0 24	0 10	0.45	0 040	0.008	3 88	8 75
*Elba, Mesaba	56 37	3 60	0 86	0 12	0.06	1 00	0 034	0 006	4.25	8.50
*Leetonia, Mesaba	54 34	2 41	0 62	0 09	0 04	0.57	0 054	0.004	6 13	11 60
*Sparta, Mesaba	55 76	7 49	0 81	0 15	0 12	0 47	0 024	0 009	2 05	9.00
*Vivian, Mesaba	38 60	35.30	1.42	1 84	0 95	0 12	0 014	0 012	1 54	3 50
†——, Clinton, N Y	44 10	12 63	5 45	...			0.650	0.230	...	2.77
‡Soft ores, Alabama	47 24	17 20	3.35	1.12	..	..			...	7.00
‡Hard ores, Alabama	37 00	13 44	3 18	16 20	..	..	0.370	0 070	12 24	0 50

Limonites.

‡——, Alabama	48 54	11 22	3 61	0 84	..	...	0.380	0.090	6.00	7.00
§Oriskany, Virginia	44 70	13 00	2 50	.		1 50	0 100	0 030	...	8.00
§Roanoke, Virginia	40 80	16.60	4.20	...	...	2 30	0.50	..	...	7.00
†——, New York	46 45	14.10	3 05	.	...	..	0.370	...	...	...
?——, Staten Island	39 72	14 19	3.59	...	..		0.059	0 301	.	12 41
‡Juniata, Penna.	43.40	18 70	5 40	0 56	0.70	0.31	0 390	0.056	...	10.37

Magnetites.

†Chateaugay, N Y	49 24	18 48	.	.	..	...		0 020	0 052	..	...
†Chateaugay, conc .	66.00	..	..	..	..	...		0 003	...	...	...
†Mineville, New York	62 10			...	...	...		1 198	...	...	...
†Hibernia, New Jersey	53.75	..	.	.	.	...		0 364	...	...	...
§Cornwall, Pa	45 86	14 07	2 29	2.88	6.37	0.54		0 022	2.20	...	8.26
§ Ditto, roasted	47 10	16 05	4 73	3 00	6 80	.		0 011	0 90	.	. .
†Cranberry, N C	.64 64	.	.		..	...		0 004	0 115		. .
†Calumet, Colorado	49 23	3 85	.	..	...	...		0 026	...	...	...

Foreign Ores

												TiO2	
‡Porman, Spain	48.95	14 04	1 31	0.74	0 24	0 85	0.041	0 271	12.53	3.55
§Bedar, Spain	52.43	4 22	0 73	2.87	0 62	1.71	0.022	0.056	7 60	9.30
§Bilbao, Spain	51 83	11.76	1 70	0 45	0 14	0 84	0 048	0 025	.	4 65
§Calaspara, Spain	61 30	4 25	1 61	1.88	2.42	0 19	0.149	0 060	...	...
§Juragua, Cuba	53 87	13 95	2 19	3 10	1 36	0 24	0 025	0 237	...	0 99
§Spanish-American, Cuba	62 85	7.66	0 58	0 51	0 45	0 15	0 028	0 147	...	0 95
——, Newfoundland	47 82	9.86	6 35	0 11	0 07	0 33	0 005	0 014	25 10	
§Wabana, Ditto	53.20	13.00	4 30	1 80	0 50	0 30	0.70	0 170	...	...

Scale and Cinders.

									Cu	
§Blue Billy, N. J	62 89	2 11	.	..		. .	0 008	0 121	0 06	...
§Roll scale, East. Pa.	71 36	2 48	0 08			0 38	0 125	..	..	2 06
§Heating cinder Ditto	52 30	25 00	2 09	9 25	0 015	0 30	0.08		..	..
§Puddle cinder. Ditto	50.75	19 88	.		.	1 19	1 57	0.14	.	0.91

* From published analyses of Lake Superior Iron Ore Association, 1904.
† From Kemp's " Ore Deposits "
‡ Phillips, " Iron Making in Alabama."
§ Private notes

FUEL

The prime object of the use of fuel in a blast furnace is the production of heat. At the same time it acts as a reducing agent, separating the iron from its oxygen. Since the fuel needed to furnish the required heat is always in excess of the requirements for reduction, it is only as a producer of heat that it demands consideration.

The materials in the hearth of a blast furnace, as in any other form of melting furnace, will melt at a rate proportionate to the rate of heat development. The development of heat will be in proportion to the rate of union between the carbon of the fuel and the oxygen of the blast. Therefore any factor which tends to facilitate this union will increase the rapidity of operation. One of the most potent factors is the character of the fuel. Any fuel which tends to impede the movement of the gases or which is not easily attacked by the blast should be avoided.

The characteristics which make a fuel desirable for use in a blast furnace are as follows:

A Well-Developed Cell-Structure.—A porous fuel will present more surface to the action of the blast than a dense one, and therefore facilitates and hastens combustion.

Firmness.—A fuel which changes its shape in the furnace, either through being crushed by the weight of accompanying materials or through softening under the action of heat, is undesirable, as the filling of the interstices of the charge with fine or pasty material impedes the current of gases and hampers combustion.

Purity.—Other conditions being equal, it is evident that the higher the fixed carbon the more efficient the fuel. The noncarbonaceous material develops no heat, but forms a slag which absorbs heat in melting. It is in the non-carbonaceous material or ash, also, that is found the phosphorus which is so deleterious to pig irons.

CONSTITUTION OF FUELS.

Solid fuel consists essentially of two parts:

(1) The combustible portion, consisting of carbon and hydrocarbons that can unite with oxygen, thus developing heat and passing away as invisible gases.

(2) The incombustible portion, which does not unite with oxygen and is left behind as a solid residue, commonly called ash. This portion consists of earthy compounds which are often difficult to fuse and have to be fluxed from the furnace.

Up to the present time only solid fuels have been used successfully in the blast furnace. Those which have been proved suitable are of three general types, namely, charcoal, raw coal and coke. Of these three only one, raw coal, is a natural fuel. The other two are produced from natural fuels, and are designed especially for use in the blast furnace. Charcoal is obtained by distilling the volatile matter from wood, thus leaving behind only the solid carbon. Coke is produced from bituminous coal by a similar operation.

Characteristics of Blast Furnace Fuels.—Each of the

three types of fuel is desirable from the standpoint of some one of the above requirements.

(1) From the standpoint of **cell-structure,** porosity and general accessibility to the blast, anthracite coal is the least desirable of the three. It is a dense substance which offers comparatively little surface to the action of the blast, and in consequence burns slowly. Coke, being made by the expulsion of volatile constituents, is naturally more porous than anthracite. It offers considerably more surface to the action of the blast, and its rate of combustion is two to two and a half times that of anthracite. Charcoal is even more porous than coke. It presents three times as great an area for contact with the blast. The calorific effect of a unit of fuel in a unit of time depends upon three factors, (1) the area of fuel exposed to the action of the oxygen, (2) the affinity of oxygen for the given form of carbon, (3) the pressure and temperature of the blast. Since for any given fuel the first two factors are fixed, it is only by control of the third that increased activity can be obtained.

(2) From the standpoint of **firmness** it is probable that charcoal is the least desirable, as it is a very friable substance and does not resist well the crushing force of the charges. Soft coal, as produced from the bituminous beds of the Pittsburgh district, is out of the question as a blast furnace fuel, since it fuses into a pasty semi-liquid mass at moderate temperatures. The anthracite coals of eastern Pennsylvania, however, are firm and strong and do not soften at any temperature. On the other hand, they decrepitate or splinter into fine particles under the influence of heat and thus become undesirable. Strong, well-made coke, such as is made from the Pittsburgh coal seam, is able to stand the crushing effect of the charges in the highest furnaces. As it has already had a baptism of fire during its preparation, it is not altered by the heat of the furnace and retains its original shape until it is burned by the blast at the tuyeres. Hence it is most desirable from this point of view.

(3) From the standpoint of **purity** and freedom from non-carbonaceous matter, charcoal unquestionably stands first. Since it is made from vegetable growths, mineral matter is present only in small quantities. The quantity of ash is small, and, in conse-

quence, there is little phosphorus. Next in order of purity is anthracite coal. When clean and well-picked, it is very high in carbon and reasonably low in ash Its content of phosphorus and sulphur is also generally fairly low. With coke, the ash, and in consequence the phosphorus and sulphur, is likely to be higher than in anthracite and much higher than in charcoal. Coke will be much higher in ash than the coal from which it is made. This is because a ton of coal is condensed during coking into about two-thirds of a ton of coke. Since the ash is not volatile, but remains with the coke, the ash of a ton of coal is concentrated into two-thirds of a ton of coke and consequently appears by analysis to be 50 per cent. higher in the coke than in the coal. According to McCreath, the phosphorus in Pittsburgh coal varies from a trace to 0.125 per cent., while that of the coke varies from a trace to 0.200 per cent. In discussing the purity of fuels, Birkinbine gives as illustration the following comparative averages:

Tr A. I M. E., VIII , p. 75.

Jour. U S. Ass'n, Char Iron Workers, III., p 361.

	Per cent P.
Charcoal	0 011
Anthracite coal	0 018
Coke	0.029

It is evident, therefore, that a fuel low in ash is of double advantage, for while it is lower in impurities, it likewise needs less flux to render the ash fusible.

In summing up the advantages of the various types of blast furnace fuel, we are led irresistibly to the conclusion that charcoal is the most desirable, coke is next, and anthracite the poorest form. Yet all are in daily use in various parts of the country, the determining factor in each case being cost alone.

In the early days of iron-smelting, charcoal was the only suitable fuel known. Wood was abundant, and the demand for iron was small. It was easy to obtain a supply of charcoal sufficient for all needs. Each locality furnished its own fuel, and the small local deposits of ore served as an ore supply adequate to the needs of the times. Every state was then a producer of iron. As the demand for iron increased, and charcoal became more scarce, a new form of fuel became imperative. About the third decade of the last century it was found that iron ore could be smelted in the blast furnace with anthracite coal as fuel. This fact gave great impetus to the iron industry in eastern Pennsylvania and vicinity, owing to its proximity to the supply of anthracite. Improvements in transportation facilities enabled this cheaper product to kill the local industry in districts remote from the coal supply.

The discovery of vast beds of coking coal in western Pennsylvania made practicable another fuel, which, if not cheaper than anthracite, is more efficient. This discovery enabled the iron industry of that locality to become, during the last quarter of the last century, the greatest in the world. With improved transportation this more efficient fuel has invaded even the home of the anthracite iron industry. There are now few furnaces in the eastern district that do not use more coke than anthracite in their charges, and many have abandoned the use of anthracite entirely The use of a given fuel is determined by local conditions. In the forest districts of the northwest, which are remote from coking coals, charcoal is still used for smelting iron on a small scale. In the vicinity of the anthracite coal regions—also remote from coking coals—varying proportions of anthracite are used to counterbalance the high cost of the distant coke. In the great region between, however, coke is the universal blast furnace fuel. It may be safely said that it is used in 90 per cent. of the pig iron production of the United States.

NATURAL FUELS.

Anthracite coal is a natural blast-furnace fuel, as it possesses properties which enable it to be used successfully without preparation. Such use is confined, however, to the larger sizes, which range from fist to head size.

Geologically, anthracite appears to be the ultimate product of the conversion of vegetable matter into coal. It contains very little volatile matter. The fixed carbon usually amounts to at least 90 per cent. It is a very compact black substance, having a brilliant sub-metallic luster. It is brittle, having an uneven conchoidal fracture, and does not soil the fingers. It is difficult to ignite and burns with a feebly luminous, smokeless flame. It decrepitates on heating and breaks up into a state of fine division.

Dry bituminous coals are used for smelting in some countries, but have no application in this country.

Coke and charcoal are fuels prepared from soft coal and wood respectively. The process consists, in both cases, of heating, out of contact with air, to such a temperature that the volatile matter is expelled without the combustion of the solid carbon. The solid carbon, together with the mineral matter comprising the ash, constitutes the blast furnace fuel.

Formerly this distillation was accomplished in heaps in the open air. Later crude kilns or ovens were used, from which the volatile matter escaped into the air and was lost. Recently great progress has been made in methods for recovering the various constituents of the volatile gases as by-products of considerable value.

Coke.—Coke is the solid residue resulting from the distillation of bituminous coal. Its manufacture is of comparatively recent date. It was first called into use in England by the increasing scarcity of charcoal during the eighteenth century, and had practically replaced charcoal there before the American Revolution.

Coke varies much in appearance, according to the coal from which it is produced and the manner of production. Good coke is of gray color and porous, yet hard and resisting. The quality is greatly affected by the temperature and duration of the process. In general, the higher the temperature and the longer the exposure the harder the coke.

Coke can be produced only from what are known as **"coking" coals.** Coking is not a universal property even of bituminous coals. A coking coal when heated to certain temperatures will swell, become pasty and emit bubbles of combustible gas, which leave behind them a solid, coherent mass of carbon. The result is the same whether the coal is in lump or powdered form. Coking does not take place at temperatures below that at which the coal undergoes decomposition. It is not the result, therefore, of simple fusion, but of chemical change. The degree to which coking may go in different coals ranges from complete fusion to mere fritting, and the line between coking and non-coking coals is not sharply defined. Indeed, mixing a portion of non-coking coal with highly fusible coal may result in perfectly satisfactory coke. The property of coking is not clearly understood. It does not appear to depend upon the composition of the coal. It may be destroyed entirely by prolonged exposure to air or to slightly elevated temperatures.

Coking.—The process of coking consists essentially of driving off at a high temperature all the volatile constituents of the coal. The volatile matter consists chiefly of hydrocarbon gases. The solid residue contains the non-volatile or " fixed " carbon, together with the ash, the phosphorus and most of the sulphur. As the coke is usually about two-thirds the weight of the coal used, it follows that all residual constituents will be higher in the coke. The following table illustrates the change due to distilling some standard coals:

Locality.	Substance.	Moisture	Volatile.	Fixed car.	Ash.	Sulphur.
Connellsville	Coal	1.25	31.27	59.79	7.16	0.53
Connellsville	Coke	0.63	1 37	85 99 ~	11.12	0 89
Connellsville	Coal	29 02		61.61	9.37	0 77
Connellsville .	Coke	1.85		87.07	11.08	0 75
Davis	Coal	23.72		63.57	7.91	0 737
Davis	Coke	1.12		88 60	10.28	0.669
Thomas	Coal	25.42		63.40	11.18	0 672
Thomas	Coke	1.20		85 45	13.35	0 665

Although by-product coking plants are springing up in various sections and the quantity of by-product coke in use is constantly increasing, the bulk of the coke used in blast furnaces is still made in beehive ovens without the recovery of by-products. This great daily waste of valuable material is due chiefly to the fact that in order to make recovery profitable the coking plant should be located near a large city or industrial establishment. At present much capital is invested in beehive ovens located at coal mines, remote from points of consumption.

Beehive Coke.—The beehive oven, as its name implies, is a dome-shaped affair, built of bricks and having a circular ground plan. It is usually about 12 feet in diameter and 6 feet 9 inches high, with a door 2½ feet square in front and an opening about 15 inches in diameter in the crown. The charge of raw coal, consisting of about 5 tons, is dropped into the oven through the opening in the crown. The charge is leveled and the door bricked up to within an inch of the top.   This space is left as an inlet for air.

As the oven is still hot from the previous charge some of the volatile gases begin to distill at once.   Then, as the mass gathers heat from the brickwork, the temperature rises above the ignition point of the gases and they begin to burn in the space above the charge.   The air to support their combustion enters the slit in the door, and the products of combustion pass out of the opening in the crown.   The burning gases raise the temperature of the

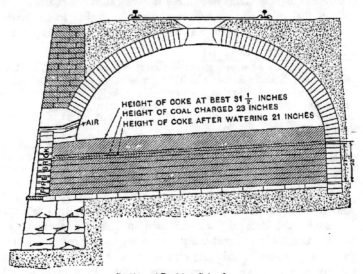

HEIGHT OF COKE AT BEST 31½ INCHES
HEIGHT OF COAL CHARGED 23 INCHES
HEIGHT OF COKE AFTER WATERING 21 INCHES

AIR

FIRE BRICK

Section of Beehive Coke Oven.

oven, and the distillation continues progressively from top to bottom of the coal.   At the temperature of the oven the coal fuses into a pasty mass.   During distillation it increases in volume, rising several inches in the oven.   The maximum swelling occurs about 3½ hours after ignition.   As the volatile matter escapes, leaving only the carbon behind, the mass again becomes solid and shrinks 20 to 25 per cent.   The air admitted during coking should not exceed the amount needed to consume the gases, as any excess will attack the solid carbon and cause a low yield of coke.   The time consumed in making furnace coke is usually 48 hours. The charge is 5 net tons and lies 23 inches deep. For making better grades, such as foundry coke, 72 hours is allowed. The charge is 6 tons and 27 inches deep. When the distillation is complete, the door is torn down and the hot coke is partly cooled by a stream of water. This causes the mass to contract and split up up all directions. It is then drawn out through the door and further cooled by water. The oven is then ready for another charge.

Coke made in this way shows a distinct columnar structure. The columns are about eighteen inches long, and represent the depth of the mass of coke. The main body of the columns is compact, showing a silvery white luster, with here and there clusters of brightly polished nodules and filaments. These are carbon which has been deposited by the decomposition of hydrocarbon gases during the period of distillation. The lower ends of the coke which come in contact with the floor of the oven do not receive such high temperatures as the upper portions and are usually darker in color and softer in texture. These "black ends," as they are called, are said to offer less resistance to abrasion. Being readily dissolved also by the carbon dioxide of the gases in the top of the furnace, they are probably carried away before they have served any useful purpose.

By-product Coke.—Coke made in retort ovens with a saving of by-products is being rapidly introduced for use in blast furnaces and foundry cupolas. This method of coking differs from the beehive method in that the operation is carried on in closed retorts so that the gases may not burn, but be saved for future use. Since the distillation begins at the retort walls, and progresses toward the middle of the mass of coal, it follows that each side distills a thickness of only half the width of the retort, which is nine inches. As a result the operation takes less than 30 hours, as compared with 48 for beehive ovens. However, since the gas escapes at the median plane of the mass, a plane of cleavage is left, and the resulting coke is only nine inches in length. Its small size and lack of luster make it appear less desirable than beehive coke. Direct comparison of the two, however, in the same furnace under similar conditions shows that retort coke is capable of carrying a heavier burden, and of producing more iron of a given quality with less stone, less fuel, and lower blast temperature, owing to its higher percentage of fixed carbon.

	Pounds ore.	Pounds fuel.	Limestone, pounds.	Blast temperature.	Output, tons.	Si.	S	Duration test, days
Kind of coke								
Beehive . . .11,310		1,885	2,250	1,014	228	1.10	0.035	4
Otto 11,575		1,787	2,050	924	233	1.06	0.039	4

Iron Age, June 11, 1903, p. 32

Retort Ovens.—There are several different types of retort coke ovens. The three which have been introduced into this country are the Otto-Hoffman, the Semet-Solvay and the Koppers. The difference between them lies chiefly in the details of construction, such as the arrangement of the heating flues, and in the preheating of the gas and air used in the flues. They are charged by overhead larries and chutes, and discharged by a mechanical pusher, which forces the retortful of coke out of the retort bodily. The water cooling then takes places outside of the ovens, and there is no corresponding loss of heat.

The **Otto-Hoffman** by-product coke retorts were first introduced by Otto in 1881. In 1883 Hoffman added regenerators for preheating the air used in the combustion of the heating gas. The retorts are built of firebrick in batteries of about 50. Each retort is about 33 feet long, 6 feet high and 16 to 24 inches wide, and is closed at each end by airtight doors. The intervening walls are 12 inches thick and contain flues for distributing the heat which performs the distillation. The covering arch has three openings for charging coal and two others for conducting away the products of distillation. The latter pass into the collecting pipe leading to the condensing plant, where they are relieved of their tar, ammonia, etc. A portion, usually one-half, of the purified gas is then returned to the supply pipe of the ovens to be distributed and burned in the flues around the retorts. The air for burning the gases is heated in the regenerators to 2,000 degrees F. The products of combustion pass up the vertical flues, along the horizontal, and down to the opposite checkers, thence to the chimney. By means of a reversing valve, the course of the gases is reversed at stated intervals. A 7 ton charge is treated in

Fulton's Treatise on Coke, 2nd Ed., p. 237.

24 to 36 hours. The yield per net ton of Pittsburgh coal is approximately as follows: Coke, 1,300 pounds; gas, 5,000 cubic feet; tar, 75 pounds; ammonic sulphate, 20 pounds.

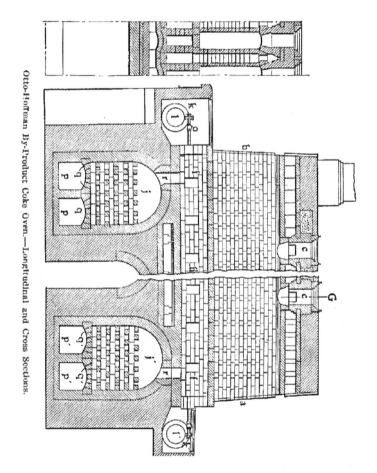

Otto-Hoffman By-Product Coke Oven.—Longitudinal and Cross Sections.

The **Semet-Solvay Retorts** were introduced in 1887. They also are built in batteries. The retorts are slightly smaller than the Otto, however, and are not provided with regenerators. They are 30 feet long, 5½ feet high, and 16½ inches wide. They have

Ibid,
p. 264. three openings for the charging of coal and one for the exit of
gas. The sides of each retort are composed of jointed horizontal
flue tiles, having walls 2¾ inches thick. The brick wall between
two adjacent retorts is 16 inches thick, exclusive of the flue tiles.
These thick walls serve to store up heat toward the end of the
period, when the charge is hot, and to give it out when a cold
charge is introduced. This makes regenerators unnecessary, and

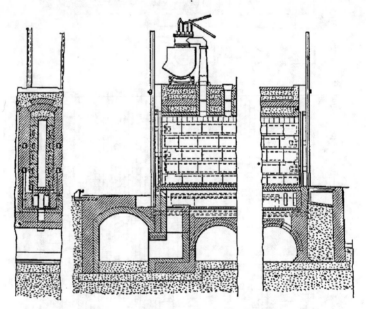

Semet-Solvay By-Product Coke Oven.—Longitudinal and Cross Sections.

the current is therefore constantly in one direction. The high
initial temperature melts the fusible constituents at the start,
thereby making stronger coke. This type of retort is especially
good for poorly coking coals. A charge of 4½ tons is coked in
18 to 26 hours.

Standard Coke.—Coals from different beds do not yield
the same quality of coke even in the same method of coking. The
Pittsburgh vein in the vicinity of Connellsville, Pa., has always
yielded coke of the first quality, it being very strong and hard, with

good cell structure and low in phosphorus. It is a little higher in ash than some other good cokes, but its other qualities have long made it the standard of comparison. Its closest competitor for general favor is made from the coals of the Pocahontas district in West Virginia. It is inferior to the Connellsville coke in physical properties, but is much lower in ash and higher in fixed carbon.

Impurities in Coal.—The chief impurities in coal, considered as a metallurgical fuel, are sulphur and phosphorus. Sulphur is always present in coal. It may exist in combination with bases as sulphates, or united with iron as pyrite, but is principally in combination with the organic constituents of the coal. The pyrite may exist in conspicuous masses or it may be disseminated through the coal as invisible particles. During the distillation a portion of the sulphur is evolved as sulphuretted hydrogen, or as a bisulphide of carbon. The proportion thus eliminated ranges up to 30 per cent of the whole.

The phosphorus occurs combined with bases as phosphates, particularly the phosphate of lime, or apatite. These phosphates are not decomposed during distillation, but are concentrated with the ash, and hence represent a higher percentage of the coke than of the coal. Coals which are high in mineral matter such as slate, pyrite, etc., are sometimes washed before coking. They are crushed to meal sizes, and jigged, whereby the quantity of ash is lessened and distributed more evenly, thus making a stronger coke.

Statistics.—The first by-product coke ovens in the United States were put into operation in Syracuse, New York, in 1893. Since then, the number has increased rapidly. They are not confined to the coal-producing or smelting districts, but are being established in manufacturing communities, where there is demand for gaseous as well as solid fuel. The comparative output for the last few years is as follows:

	1905	1906.	1907.	1908
Number ovens existing	3,159	3,603	3,892	4,007
Coke produced, net tons	3,462,348	4,558,127	5,607,899	4,201,226
Average product per oven, net tons	1,159	1,356	1,472	1,142
Per cent. total coke production	10 7	12 6	13.7	16 1
Total coal used, net tons	4,628,981	6,192 086	7,506 174	5,690,058
Average yield, per cent	75	73 6	75.0	73.7

Coke is made in all of the seven bituminous coal fields of the United States. By far the greatest amount comes from the Appalachian district, which includes such important iron-producing states as Pennsylvania, Ohio, Virginia, West Virginia, Kentucky, Tennessee, Georgia and Alabama.

Charcoal.—When wood is heated without contact of air, it breaks up into certain volatile products and a fixed carbonaceous residue known as charcoal. The charcoal retains the structure of the original wood, is black in color, very light and extremely porous. Good charcoal gives a sonorous ring when struck, and while it breaks easily, is not readily crushed by ordinary pressure. When ignited it burns without flame.

While charcoal retains almost the bulk of the original wood, its specific gravity is very low. As a result the average yield of charcoal is 20 to 25 per cent. of the wood Of this the ash rarely averages over about 3 per cent. Charcoal made at low temperatures and of soft wood ignites most readily. That which is burned at high temperatures ignites with difficulty.

Charcoal Making.—Charcoal, which was formerly burned in heaps or kilns without recovery of the by-products, is now usually made in retorts with recovery. It is made from various kinds of wood, and its properties vary accordingly. Its weight varies widely from 23.5 pounds per bushel when made from black oak, to 14 pounds per bushel when made from alder. A bushel of mixed charcoal weighs usually about 20 pounds. It is convenient to assume that 100 bushels make a net ton. The following are some average analyses of charcoal made from different kinds of wood:

Kind of wood.	Fixed carbon. Per cent.	Volatile matter. Per cent	Moisture. Per cent.	Ash. Per cent.	P. Per cent
Pine	89 52	4.46	3.15	0 87	0.014
Ash	89 19	5.63	4 08	1 15	.. .
Spruce	87 23	6 08	5.77	0.92	0.016
Birch	86 50	7.17	5 61	0.65	0.026
Willow	86 04	4.23	5.93	3 80	....
Fir	73 60	10.30	5 52	1 58	....
Alder	81 50	8.40	6 40	3 70	....

Jour. U S. Ass'n Charcoal Iron Workers, V., p 53.

From these analyses it appears that during the distillation the volatile constituents are never wholly expelled; that charcoal absorbs moisture amounting usually to at least 5 per cent. of its weight and that the ash and phosphorus are both very low as compared with coal and coke.

The average analysis of wood is approximately as follows:

	Per cent.
Fixed carbon	51.6
Hydrogen	6.5
Oxygen	41.8
Nitrogen	0.1

Iron Age, July 28, 1898, p. 6

It is apparent that under the most favorable conditions there must be more waste and less recovery in making charcoal than in making coke, since the volatile constituents are in larger proportion. For this and other reasons the use of kilns and closed retorts was early adopted in charcoal making.

The making of charcoal in retorts at the Algoma Steel Company plant at Sault Ste. Marie is conducted as follows: The retorts consist of rectangular, horizontal shells 46 feet long, $6\frac{1}{4}$ feet wide, and $8\frac{1}{3}$ feet high, made of $\frac{3}{8}$-inch plates. They are riveted together with calked seams, and fitted at the ends with doors, which are capable of air-tight sealing. They are set in brick walls like a boiler, and have a fireplace at each end with flues leading to a stack so arranged as to heat the retorts evenly. The wood is split and seasoned, then loaded on trucks and run into the retort, where it is subjected to heat for 24 hours, 18 or 20 of which it is at the full temperature of carbonization. The distillation products are led to the by-product plant for the recovery of alcohol, acetic acid, tar, etc, and the gas is sent to boilers to raise steam. After the distillation is complete, the trucks are pushed into the first cooler. The coolers are duplicates of the retorts. After a stay of 24 hours, the trucks pass to the second cooler for 24 hours, when the charcoal is ready for the blast furnace. This system obviates all rehandling of materials, as the wood is not touched after it is placed on the trucks until as charcoal it is charged into the furnace.

Iron Age, Jan. 28, 1904.

The following is a comparison by Sjöstedt of the value of a cord of wood carbonized under different systems:

	Bushels charcoal.	Gallons wood alcohol.	Pounds calcie acetate.	Gallons spirits turp.	Tar.	Value per cord
In meilers, " coaling under dust "	33-35	0	0	0	0	$2 04
In kilns without recovery 43-45		0	0	0	0	2.64
In kilns with recovery :						
Mixed hardwood........	43-45	2-2½	50-100	..	3	4 04
Southern pine........	43-45	1	50-75	5	5	4 98
In retorts :						
Mixed hardwood.........50-52		8-12	150-200	..	5	7 65
Southern pine 50-54		1½-2	120	10	10	7 37

Ibid.

The advantage of retorts is that the by-products are preserved uncontaminated, and cheap fuel may be used for distillation.

FUEL CONSUMPTION IN THE UNITED STATES.

Until 1855 the chief blast furnace fuel in America was charcoal. In that year the output of anthracite pig for the first time exceeded that of charcoal pig. In 1869 the production of coke iron first exceeded that of charcoal iron, and in 1875 coke also surpassed anthracite as a metallurgical fuel.

The first statistics concerning the production of coke date from 1883, when 3,338,300 net tons were made. The comparative production for the past three years in the United States was as follows:

	1906	1907	1908
In Pennsylvania, net tons	23,060,511	26,513,214	15,511,634
In Alabama, net tons	3,034,501	3,021,794	2,362,666
In West Virginia, net tons. ...	3,713,514	4,112,896	2,637,123
In Virginia, net tons............	1,577,659	1,545,280	1,162,051
In other States, net tons............	3,015,032	5,586,380	4,360,044
Total in United States, net tons.	36,401,217	40,779,564	26,033,518
Total coal used, net tons 55,716 371		61,946,100	39,440,837
Average yield, per cent........ . ..	65 3	65 8	66.0
Number ovens existing	93,901	99,680	101,218
Number active....................	88,596	94,746	88,298
Average product per active oven, net tons	410 9	430 4	294.8

It may be observed that in 1907 the yield of beehive ovens was 60.7 per cent. as compared with 75 per cent. for the by-product ovens.

FUEL ANALYSES.

	H_2O	Volatile matter	Fixed carbon	Ash	S	P	SiO_2	Al_2O_3	Bases	Fe
Connellsville. Pa., coke (aver. 200 cars)	0.32	1.70	86.50	11.50	0.75	0.020	5 30	3.00	4.30	2.60
Pocahontas, W Va., coke (av. 405 cars).	2.31		90 62	6 97	0.70	...	...	...	...	...
New River, W Va, coke		2 82	90.52	6 36	0 70	...	...	...	...	...
Davis, W Va , coke.	0.33	1 20	87 97	10 50	0 55	0.034	4 27	3 89	...	1 40
Derry, Pa , coke		2 50	81 60	14 70	1 20	...	.	...	...	...
Webster, Pa , coke	1.10		88 40	10.50	0 95	0 03	4.80	3.50	...	1.85
Klondike, Pa , coke		2 50	84 50	12.00	1.00	...	...	...	...	...
Loyal Hanna, Pa , coke.		2 45	81 20	15 00	1 35	.	...	...	...	...
Mountain, Pa , coke	0 50	1 10	87 85	11 53	1 00	0 018		. .	...	...
Fairmont, W.Va , coke	0 30	0 97	87 43	11.30	1.35	0.020	4 16	2 53	...	2 45
Camden by-product, Otto Hoffman	0.40	1 20	84 60	13 80	0 88	0 030	5 70	4.50	...	1 55
Anthracite, Pa		8 80	78.00	13 00	0 70	...	. .	...	...	...

VALUATION OF FUELS.

Since a fuel is primarily intended to ·furnish heat, it is evident that its valuation must be based upon the available units of carbon which it contains. The available carbon is that which is left after deducting from the fixed carbon the amount necessary to melt the slag formed by the ash and sulphur of the fuel and the stone needed to flux them. In order to estimate the approximate amount of stone needed, it is convenient and sufficiently accurate to consider that the ash requires double its weight of stone, and the sulphur three and one-half times its weight. The weight of the ash, plus the weight of the sulphur, plus the weight of the slag-forming portion of the calculated stone, will give the weight of the slag formed by the fuel. If the slag requires 25 per cent. of its weight of carbon to melt it, then 25 per cent of the weight of slag deducted from the total fixed carbon will leave the available carbon. For example, a coke, having the following analysis:

Per cent.

Fixed carbon 84 40
Ash 12 00
Sulphur 1.15

will require $(2 \times 12) + (3.5 \times 1.15) = 28$ per cent. stone. If the stone contains 43 per cent CO_2 it will contribute 57 per cent. to the slag. Hence the weight of slag per 100 pounds fuel will be $12 + 1.15 + (28 \times .57) = 29.11$ pounds. $29.11 \times .25 = 7.28$, the carbon used by its own slag. $84.40 - 7.28 = 77.12$, available carbon.

The relative powers as heat producing agents, of fuels, and consequently their relative values to a furnace man, are in proportion to their respective amounts of available carbon. To find their respective cash values, it is only necessary to estimate their relative costs, based on the amounts of available carbon, and the price paid per ton, and then to deduct the cost of the stone needed to flux the ash and sulphur. In comparing the smelting powers of various fuels, the only safe basis of comparison is that of the available carbon.

FLUXES.

The function of a flux is, as the name implies, to facilitate flow. All matter which is infusible or fusible with difficulty at furnace temperatures must have its composition so modified that the fusibility will be increased. A suitable flux cannot be selected without a knowledge of the composition of the material to be fluxed. In general it may be stated that if the matter to be fluxed be basic, such as lime, magnesia, or similar alkaline matter, the flux must be acid in nature, such as silica, alumina, etc. On the other hand, acid gangues, such as silica, alumina, etc., require basic fluxes. As a rule gangues are acid in their nature. The acidity may be partly neutralized by the presence of basic matter. There is very seldom sufficient basic matter to completely neutralize the acid components. As a result, we find that fluxes are almost invariably basic in character.

Any of the metallic oxides may act as flux for silica and alumina, but the cheapest, and hence the one most generally used, is lime (CaO). The lime is generally introduced into the furnace in the condition of calcic carbonate ($CaCO_3$), usually in the form of limestone, though sometimes as oyster shells.

When the limestone which has been charged into the furnace has descended to the point where the temperature is above $1100°$ F., it begins to decompose in accordance with the following reaction:

$$CaCO_3 = CaO + CO_2.$$

The CaO thus liberated continues to descend with the charge

until, in the zone of fusion, it unites with the gangue to form the slag. The CO_2, however, joins the current of gases and passes upward. If it encounters on its way any coke which is at sufficiently high temperature, it will attack it thus,

$$CO_2 + C = 2CO$$

and the carbon thus dissolved will be carried out of the furnace, without having developed any useful heat. It would appear advantageous, therefore, to decompose the limestone before charging it, since only the CaO is needed and the CO_2 steals coke. Experiments indicate, however, that the advantages of using caustic lime instead of limestone are not marked. This is well shown by excerpts from experiments by Sir Lowthian Bell.

Furnace.	Flux.	Coke, pounds	Weight flux pounds	Ore, pounds.	Iron, tons. per week.	Grade	Time, weeks	
No. 11.	Limestone 2,182		1,176	4,722	451	3.3	19	Inst. Jour, 1894, II., p. 49
No 11	Lime 2,009		862	4,734	497	3.3	17	
No 12.	Limestone ....2,201·		1,187	4,762	459	3 3	19	
No 12.	Lime 2,014		857	4,740	481	3.3	17	

From the above it appears that with caustic lime, less fuel is needed and more iron is made, of the same quality from the same burden. The additional cost of calcination, however, about offsets the gain.

Experiments by Cochrane show very similar results.

Time.	Lime, pounds.	Lime-stone.	Fuel, pounds.	Ore, pounds.	Blast temper-ature	Blast press., pounds.	Out-put, tons.	$\frac{CO}{CO_2}$	Temp. waste gases	Wt. waste gases	
6 months....	0	1,391	2,122	7,238	1,474	5.04	3,584	1·77	607	125 62	Inst. Jour, 1898, II , p
4 months..	.599	504	2,019	7,038	1,489	4 97	3,625	1 94	589	120 02	

On half lime there is a decrease of fuel and increase of output with better extraction and cooler top. The benefits which would naturally be expected from the use of caustic lime as a flux are neutralized by its tendency to absorb CO_2 from the escaping gases. This reaction partly undoes the results of calcination, as CO_2 is absorbed and carried down with the charge, only to be released and, later to dissolve some of the coke, thereby decreasing the expected economy.

The limestone is charged into a furnace crushed to pass 4 to 6 in. rings. By the time it has reached the zone of fusion it has been practically calcined, and only the CaO remains. It is the duty of the CaO to unite with any free silica and alumina and to form slag. As stated above, some of the silica and alumina may already be in combination with an equivalent of bases which were present in the original gangue of the ore. The lime is added only to take care of the surplus of acids. On the other hand, few limestones are so pure that they do not contain a small percentage of silica and alumina, and their demand for bases must be satisfied before there will be any bases available to flux the gangue and ash. As the amount of bases in the slag of a blast furnace is usually about equal to the acids, it is evident that the presence of acids in limestone rapidly decreases its efficiency, thus making it imperative that pure stone be used. A suitable stone should contain at least 95 per cent. pure carbonate.

In addition to combining with the silica and alumina, the CaO should seize and carry into the slag as much as possible of the sulphur. Ten per cent. of the sulphur in a charge, if allowed to enter the iron, is usually enough to make pig iron unsuitable for most purposes It is imperative, therefore, that the greater part of the sulphur should be removed. This duty devolves upon the lime, which unites with the sulphur, forming calcic sulphide (CaS), and enters the slag.

In some localities pure limestone is not available and it is necessary to resort to the magnesian variety, which is called dolomite. True dolomite contains one atom of magnesium to every one of calcium, and may be represented by the symbol $MgCa(CO_3)_2$. Not all magnesian limestones attain to the full condition of dolomite. The proportion of magnesium varies from traces to equivalent parts.

The presence of magnesium in small quantities makes no appreciable difference in the behavior of the flux. When associated with calcium in considerable proportion, however, it tends to make the slag less refractory than the calcium alone, since double silicates are more fusible than silicates of a single base.

It has been claimed by some that magnesian limestone does not hold sulphur in the slag as well as pure limestone. F. Firmstone states that the substitution of dolomite for limestone in the furnace at Glendon, Pa., made the sulphur in the iron lower and

M. E.,
p 498.

more regular. At the same time the furnace worked better. It is stated also by E. A. Uehling that the substitution of dolomite at Birmingham, Ala., for the limestone up to three fourths of the total gave only good results. The sulphur in the coke was above the average, yet the iron was of good quality and the furnaces worked better than on limestone. W. B. Phillips states that in Alabama "the use of dolomite is a decided advantage, especially in the elimination of sulphur." It is not apparent whether this result was due to the presence of magnesium or to the absence of silica, since the dolomite was exceptionally pure, as shown by the following analysis of the limestone and dolomite.

Iron Age,
July 19, 1894

"Iron Making
in Alabama,"
Geol. Sur
Ala , p. 63

	SiO_2.	R_2O_3	$CaCO_3$.	$MgCO_3$.
Limestone	4.00	1 00	94 60	...
Dolomite	1.50	1.00	54.00	43 00

Relative Values of Fluxes.—The relative values of fluxes evidently depend upon the power of a given weight to flux refractory constituents. This quantity is manifestly in proportion to the quantity of available bases present in the flux.

The Available Base.—By the available base of a flux is meant the base that is left free to unite with the gangue and ash. It consists of that portion of the stone which remains after there has been deducted the CO_2, the acid impurities and the portion of the base which is necessary to flux the acid constituents. For example, if a limestone having 96.15 per cent. $CaCO_3$ and 3.85 per cent. acids, were to be used as a flux in a slag having acids equal to bases, the available base can be found as follows, remembering that CaO is 56 per cent. of $CaCO_3$:

$$96.15 \times 0.56 = 53.85 = \text{total base.}$$
$$3.85 = \text{base needed to neutralize acids}$$
$$50.00 = \text{available base.}$$

The available base is the only safe basis of comparison of the relative value of various fluxes.

ANALYSES OF FLUXES.

Limestones and dolomites.	SiO_2.	R_2O_3.	$CaCO_3$.	$MgCO_3$.	S	P.
Annville, Pa.	0 64	0 72	96 78	1 23	0.012	0.004
Wrightsville, Pa.	1.44	2.12	82.55	13 50	0 032	0.020
Maryland	0 78	0 86	97.48	1 21	...	...
Kelley Island, Lake Erie	1 15	1.04	82.41	15 64	...	...
Avondale, Pa. (white)	1 50	1.86	66.50	30 40	0 004	0 005
Avondale, Pa. (gray)	3 94	2.80	59 51	33.26	0.132	0.019
Buena Vista, Va	1.34	0.92	54 50	Tr.	...	...
Buena Vista, Va.	0 75	1.08	29 80	21 20	...	...
Birmingham, Ala.	1.75	1.20	81 70	10 00	...	...

CHAPTER II.

DESCRIPTION OF PLANT.

Introductory.—From the small beginnings of a century ago, the blast furnace plant has developed into a gigantic affair with truly formidable equipment The various members which go to make up such a plant may be profitably enumerated before proceeding to a description of them.

The process of smelting iron in the blast furnace consists essentially of charging a mixture of fuel, ore and flux into the top of the furnace, and simultaneously blowing in a current of heated atmospheric air at the bottom The air burns the fuel, forming heat for the chemical reactions, and for melting the products; the gases formed by this combustion remove the oxygen from the ore, thereby reducing it to metallic form, the flux renders fluid the earthy materials The gaseous products of the operation pass out of the top of the furnace, while the liquid products, cast iron and slag, are tapped at the bottom. The escaping gases are combustible, and therefore are conducted through pipes to boilers and stoves where they perform the useful services of heating the blast and raising steam or operating internal combustion engines

It is evident, therefore, that several factors enter into the composition of a complete smelting unit The central feature is the furnace with its hoists and skips for the handling of charges, and its ladles and pig-machines for handling the products Quite as essential are the blowing engines which drive the blast to the furnace, and the series of hot blast stoves, which heat the blast on its way Not less important are the boiler plants which furnish the power to drive the blowing engines, the gas engines operating electric motors or blowing engines, and the pumps which supply the vast quantities of water needed for cooling purposes and for power development.

THE FURNACE.

A blast furnace consists essentially of an enclosed space or shaft for reducing and melting materials, and a crucible for collecting the molten products of the operation. The crucible is cylindrical. The shaft has the shape of two conical frustrums placed base to base, and surmounts the crucible. The greatest width of shaft will evidently be at some point between the crucible and the top of the shaft. By such an arrangement the pathway presented to the materials descending the shaft constantly grows

View of Hearth of Blast Furnace.

larger, thereby facilitating their descent while they are solid. The space contracts as the materials approach the zone of fusion, and the converging walls furnish support for the lessening bulk. The crucible serves to hold the liquid iron and slag until the accumulation is sufficient to be drawn off.

Such in brief are the essential features of the furnace. The crucible is more often termed the "hearth." The point of greatest diameter in the shaft is called the "bosh." The region en-

closed by the sloping walls which connect the hearth and the bosh
is termed "the boshes" of the furnace. The tapering walls
reaching upward from the bosh are called the "inwalls." The top
of the furnace is closed by the "bell and hopper," which rests on

View of Hot-Blast Stove Connections.

the top of the inwalls. The portion of the shaft just below the
hopper is called the "stockline."

Though furnaces are built in many sizes, the dimensions of
these various parts are generally proportional. The maximum
sizes which have been so far attempted are as follows: Total

height, 106 feet; bosh diameter, 24 feet; hearth diameter, 17 feet;
stockline, 18 feet

FURNACE CONSTRUCTION.

From the point of view of construction, the blast furnace may
be looked upon as being made up of a steel shell and a brick
lining. The shell of the shaft above the boshes is always constructed independently of the shell of the parts below. It consists usually of steel plates ⅝ to ¾ inch thick, riveted together
and supported by a mantle resting upon columns. These columns
are usually 8 or 10 in number, and from 15 feet to 25 feet high.
They rest upon the main foundation of the furnace and form a
circle, which, in large furnaces, is about 30 feet in diameter.
Within this circle the hearth is located. Its foundation must be
of heavy firebrick construction. The hearth itself is made of refractory firebricks, and is surrounded and supported by the hearth
jacket, which is usually composed of heavy segmental iron or
steel castings or of heavy riveted steel plates. The cast jackets
are usually cooled by water flowing over troughs, or by coils of
wrought iron pipe cast in the jackets. The riveted jackets are
cooled by external water sprays.

Bosh Construction.—There are two general methods of con-
structing the bosh walls. The choice between them depends upon
the method of water cooling to be adopted. The simpler form consists of a riveted plate jacket, similar to the shaft jacket, which
is protected by a thin lining of firebricks, generally not more than
9 to 13½ inches in thickness. The system of cooling consists
either of sprays of water directed against the bosh jacket from all
sides, or of a spiral trough winding about the boshes and kept
full of running water.

The more usual method of bosh construction, however, is to
make a thick bosh wall of fire bricks, and insert at intervals bronze
cooling plates. Cooling plates are hollow bronze boxes tapped with
1¼-inch inlet, and 1-inch or less outlet pipes, so that they may be
kept full of running water. They are usually wedge-shaped,
ranging from 1 to 3 feet wide at the nose, and as long as the wall
is thick. They are 4½ inches high, with a flat bottom and an
arched top which tapers toward the nose. In order to save bronze

they are sometimes made half length and provided with a cast
iron box open at each end, into which they fit. These boxes support the bosh wall, and facilitate the removal of the plates   Adjacent plates are usually not over 4½ inches apart and the vertical
distance between rows ranges from 1 to 2 feet These bosh plates
are inserted at intervals from the crucible to the mantle   Above
the mantle two or three rows of iron coolers are sometimes used
to protect the top of the boshes. Such bosh construction, consisting of frequent alternations of brickwork and cooling plates
over which the brickwork is carried by very flat arches, needs support, and is therefore always reinforced by heavy bands of iron
encircling the furnace between the successive rows of plates.

Bosh Plates.—Bosh plates were first used by Hunt in the
seventies. They consisted then of a single coil of wrought iron pipe
embedded in an iron casting. Cast iron melts readily and leaves
the iron pipe exposed and subjected to abrasion. Owing to better conducting qualities bronze or copper coolers are more efficient and are now in universal use. The **Fronheiser** plate was
introduced first at Cambria. It was a hollow bronze box about
a foot square and perhaps 20 inches long, tapering slightly toward
the inner end. This was set in the brick work, and kept full of
running water. The **Scott** plate is flatter in shape. It is 1 to 2
feet wide and only 4½ inches high, with an arched top, and a
slight taper toward the nose. The entering water is led directly
to the nose of the plate, and is allowed to find its way back through
a series of baffles which divide the space in the plate The
Gayley plate is of similar shape, but has a water space comprising only about 10 inches of depth from the nose. The
Kennedy plate differs from the Gayley in that it has two water
ways cast in metal instead of a single large space The present
tendency is to use small plates of the Scott type.

Inwall Construction.—Above the bosh of the furnace the
temperature is not so high as to demand any system for cooling. The causes for wear on the lining of the inwalls are the
abrasion due to the friction of the moving particles of solid stock,
and the erosion of the gases The effect of moving stock is min-
imized, however, by the batter of the walls It is most marked
at the stockline where the charges strike the wall as they slide off

. M. E,
, p 102.

the distributing bell. This zone is often protected by a shield of steel plates or several rows of cast iron plates set in the brickwork. The erosion of the gases appears to be more chemical than physical in its nature. It is usually greatest above the bosh, about two-thirds way down the inwalls.

Lining.—The lining of the furnace is usually composed of 9-inch and 13½-inch firebricks. Except in very large furnaces, the inwalls are usually 27 inches thick, and are composed of fine-grained, hard clay bricks, designed to resist abrasion. The " Woodland " brick, made by the Harbison-Walker Company for this purpose has the following approximate composition:

	Per cent
SiO_2	60.65
Al_2O_3	33.81
Fe_2O_3	2.14
CaO	0.41
MgO	0.33

The lining of the boshes, when bosh plates are used, is also usually 27 inches thick, but when surface-cooled is only 9 inches or 13½ inches. The lining of the tuyere zone is seldom less than 27 inches and the crucible walls are 31½ inches or more. The hearth and bosh need a more refractory brick than the inwalls, to endure the higher temperatures. A coarse-grained, softer brick is generally used. Such a brick, furnished by the Harbison-Walker Company analyses as follows:

	Per cent
SiO_2	54.82
Al_2O_3	38.12
Fe_2O_3	3.36
CaO	1.35
MgO	0.16

Brick is very soluble in furnace slag, and would not long protect the bosh and crucible jackets from the molten material, if it were not that early in the operation of a furnace, a carbonaceous deposit replaces part of the brick in crucible and bosh walls, and offers complete resistance to corrosion. Such a deposit develops best with a hot and basic slag. Carbonized bricks usually show a composition within the following limits:

	Per cent.	
C	.25-50	*Ibid.*
CaO	.15-30	
SiO_2	.20-45	

FURNACE OPENINGS.

Tapping Hole.—The furnace walls must be pierced for the admission of the blast and the egress of products. The tapping hole for the removal of molten iron must be near the level of the hearth bottom. It is always at the front of the furnace, and usually consists simply of a passage 8 or 10 inches square left through the brickwork. It is stopped with clay, and owing to its exposure to very intense heat during the tapping of the iron and the danger of explosion when iron comes in contact with leaking

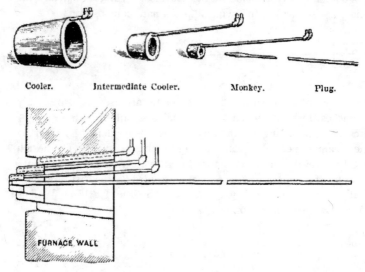

Cooler.   Intermediate Cooler.   Monkey.   Plug.

Composition of Cinder Notch System.

water, attempts are seldom made to protect the opening from heat. The other openings, however, are always water-cooled.

Cinder Notch.—The cinder notch is the opening designed for the removal of the accumulated cinder or slag. There is usually but one, although the latest construction sometimes provides two or three. Generally, however, only one is used at a time. Its location is usually about 90 degrees from the tap-hole. When there are two, they are 180 degrees apart. Since this opening is intended for the removal of only that cinder which accumulates between the casts of iron, its location must be above the point to which the iron rises, as the time of the casting is governed by the height of the cinder notch. Its height above the hearth ranges in different sized furnaces from 3½ to 5½ feet, according to the quantity of iron the crucible is designed to hold.

The cinder notch is protected, and the flushing of cinder facilitated by a series of water-coolers. In the opening through the brickwork, which is about two feet in diameter, is first placed a **cooler** which is tamped tight with clay. This cooler is in shape a hollow frustrum of a cone, and may be either a coil of iron pipe, covered by cast iron, through which water flows, or a hollow bronze casting which can be kept filled with water. The opening through the cooler is reduced to six inches by means of a shorter bronze cooler called the **intermediate cooler.** Within the intermediate cooler is a still smaller cooler, called sometimes the **"monkey,"** which reduces the opening to about two inches. The latter opening is closed by a tapering iron **plug** with a long handle. Flushing is accomplished by simply withdrawing the plug, which is again inserted when the cinder stops running. This arrangement is impossible in the openings from which molten iron flows, as the iron attacks the bronze and causes it to leak. On the other hand, a brick cinder notch would be cut out rapidly by the slag, which, however, has no effect upon the bronze.

Tuyeres.—The openings for blast admission are called **tuyeres.** They usually occupy a plane 2½ to 3 feet above the cinder notch, and determine the maximum height to which cinder may rise. The number of tuyeres in a furnace may range from 6 to 24, but 12 or 16 are most common, with recent preference for one to each foot of hearth diameter. In the tuyere opening is placed a **cooler** very similar to the cinder notch cooler. It is usually of bronze, and is set tight in clay. The cooler is flush with the brickwork, both within and without the wall of the hearth. The tuyere is inserted in the cooler and usually extends several inches beyond it. It is of copper or bronze, hollow and water-cooled like the cooler, and not unlike it in shape, though much smaller. Its external diameter must be such as to fit snugly in the nose of the cooler, to prevent leakage Tuyeres present openings ranging usually from 4 to 7 inches. This opening, together with about a

2-inch water space all round and bronze walls ¼ inch thick gives
a necessary external diameter of 10 to 12 inches. The water space
in the tuyere and also in the cooler is tapped for two 1¼-inch
water pipes, through which flows a constant and copious stream
of water. In spite of this fact, the bronze is frequently "burned"

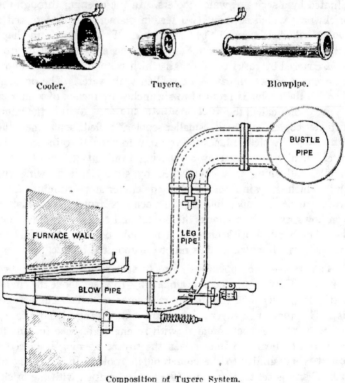

Composition of Tuyere System.

through, and leakage of water occurs, thus necessitating a change.
This accident happens far more frequently to the tuyere than to
the cooler, since it extends beyond the cooler and is consequently
more exposed. The water pressure should be about 25 pounds to
insure good circulation. Sometimes the system becomes clogged
and hydraulic or steam pressure should be at hand to force out

obstructions promptly, or the tuyere will be burned at the point of stoppage.

The blast is transmitted to the tuyere by means of a horizontal iron pipe, known as the **blow-** or **belly-pipe.** The blow-pipes are usually 4 or 5 feet long, and slightly tapered. They are turned with a slight ball at the end to correspond with a slight socket on the tuyere, in order to facilitate the connection, as in a universal joint. The joint is purely a contact union, the ball and socket being used in place of any attempt at packing.

Connecting the blow-pipe with the distributing blast pipe is a brick-lined cast iron pipe, known as the **leg-pipe, goose-neck** or **tuyere-stock.** This pipe is usually hung by lugs to eye bolts so that it can swing backward, thereby releasing the blow-pipe when it is desirable to remove it. When in place, the tuyere-stock is held firmly against the end of the blow-pipe by means of a heavy spiral spring and connecting rod, reaching from a lug on the tuyere stock to a corresponding anchorage on the hearth jacket. The longitudinal pressure so developed furnishes the sole support of the blow-pipe.

In the back surface of the tuyere-stock, in line with the blow-pipe and tuyere is a 1½-inch hole, through which a bar may be thrust to clear the tuyere, without removing the blowpipe. When the blast is on, this hole is closed by a latch, through which is a smaller opening closed by blue glass. This opening, which is known as the **peep-hole,** enables one to look through the tuyere into the furnace, and thereby to get direct evidence as to hearth conditions.

The blast-distributing pipe, usually called the **bustle-pipe,** encircles the furnace at a height about 10 feet from the floor, and connects with each tuyere. It consists of riveted steel plates and has a 9-inch lining of firebricks. Its diameter varies with the quantity of air which it must carry, and is usually 4 or 5 feet. Into it leads the **hot-blast main** which conducts the blast from the stoves.

TOP ARRANGEMENTS.

Besides the openings to admit the blast and to discharge the molten products there is ordinarily no opening into the furnace except at the top, where the materials enter and the gases escape. The top of the furnace was formerly left open, and the escaping gases burned with a constant flame. When it was discovered that these gases could be used to heat the blast and to generate power, the bell-and-hopper device was adopted to prevent their escape. A single bell accomplishes this object, except when it is lowered to dump a charge. Of late years, a double bell system has been widely adopted to prevent all waste of gas. Outlet for the gas is provided at one or several points beneath the bell. These outlets lead into a large tube several feet in diameter, known as the "down comer" or "downtake," which conducts the gases to the boilers and stoves. In some furnaces there are distributed about the top weighted doors, known as "explosion doors," which are designed to relieve the pressure caused by occasional explosions in the top of the furnace.

Bell and Hopper.—The bell and hopper together form an annular V-shaped depression into which the stock may be dumped. When the bell is lowered the whole charge slides into the furnace. In order properly to distribute the stock, it is found by experience that about two feet clearance all round the bell gives best results. The bell is usually cast iron, in one or two pieces, and of such a slope, usually 45 degrees, as to admit of its readily clearing itself of its burden. The hopper consists of two pieces, the hopper and the extension-ring. They are also of cast iron and are usually made in segments. The extension-ring is suspended from a flange on the bottom of the hopper, where it must make a gas-tight joint, and yet be readily removed to permit the changing of bells. When in place it serves to form the joint between the bell and the hopper. The bell is usually suspended centrally from a counterbalanced lever, which is controlled and operated by means of an air or steam cylinder. Usually the bell is kept closed by this pressure. When the pressure is released the bell descends from the weight of the charge upon it.

With the introduction of the second bell it is necessary to enclose the first bell and hopper in a casing. The second bell closes the opening in the top of the casing. Since both bells must be suspended concentrically and yet be operated independently, it is necessary that one be hung on a sleeve surrounding the rod which suspends the other. The lower and larger bell is usually suspended on the rod and the upper bell on the sleeve. Like the single bell, they are operated by horizontal levers controlled by air or steam cylinders.

The Bleeder.—Leading out of the top of the downtake is a short vertical pipe of small diameter, called the "bleeder." It allows surplus gas to escape from the furnace whenever necessary. It is provided with a valve on top, which may be controlled from the ground.

Downtake and Dustcatcher.—The gases which are generated during iron smelting are highly combustible, and are used as fuel to raise steam and to heat the blast. At present, however, they are being largely applied, particularly in Europe, to furnish power by direct use in gas engines. The gases are taken from the furnace beneath the bell through outlets, usually ranging in number from one to four, which converge into a single large pipe called the **downtake.** As the current of gases comes over into the downtake with considerable velocity, it tends to carry along fine particles from the charge. If this dirt is allowed to go to the combustion chambers of the stoves and boilers it accumulates there and makes frequent cleaning necessary. Proper cleaning necessitates the cooling of the stoves and boilers, which causes not only considerable loss of time and of accumulated heat, but makes it necessary to have extra stoves and boilers. Moreover, the alternate heating and cooling of the brickwork sets up strains, which tend to disrupt it. The cleaning is never thorough, since fine particles adhere to the flue walls and, owing to their lack of conductivity, reduce the efficiency of the heating surfaces.

The purpose of the **dustcatcher** is to remove as much as possible of this dirt. It is virtually an enlargement in the downtake, and depends for its action upon the principle that lessening the velocity of the current will give opportunity for solid particles to settle by gravity. It is evident that the efficiency of a dustcatcher will be in proportion to its size, since the larger the cross section of a conduit the slower the rate of motion of a given volume of gases. This fact has received more recognition in later than in former construction. The dustcatchers of the Lackawanna Steel Company's new furnaces at Buffalo are 32 feet in diameter. The rate of motion of gases in the downtake should not exceed 30 feet per second. A dustcatcher having a diameter four times greater than the downtake would have a cross section sixteen times greater, which would reduce the velocity of the gases to about 2 feet per second. This velocity will not ordinarily carry more than 3½ grains of solid matter per cubic foot. Such gas may be used in stoves and boilers, but is still out of the question for gas engines.

Gas Cleaning.—In order to get the best results in stoves and boilers the gases should also be at least partially washed. The

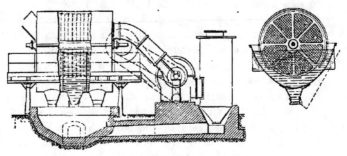

The Bian Gas Washer.

several types of gas washers may be classified under three general heads, as stationary scrubbers, slowly revolving washers, and rapidly revolving washers or centrifugal machines.

The **stationary scrubbers** consist generally of vertical chambers containing sprays of water, and sometimes provided at intervals with some type of porous filter, which serves to hold the water in a finely divided condition and at the same time to split up the current of gases and bring the two into intimate contact. The Zschocke cleaner is a good example of this type. It brings the dust down to about 0.65 grains per cubic foot.

Iron Age, May 18, 1905, p. 1586.

Ibid, July 14, 1904, p. 16.

Revolving cleaners consist essentially of stationary horizontal cylinders, having through their centers slowly revolving shafts, carrying perforated discs which are half submerged in water. The gases flow above the water and are forced through the perforations which are charged with water. In this way fresh water is constantly being presented to the gases. The Bian apparatus is said to wash 40,000 cubic feet of gas per minute with expenditure of about 35 H. P.

Such methods of washing have their defects. They are not sufficiently thorough for gas engine work, and they cool the gases more than is desirable for consumption in stoves and boilers. Less thorough methods answer all requirements for the latter purpose. The Steece washer, for example, which gives satisfaction

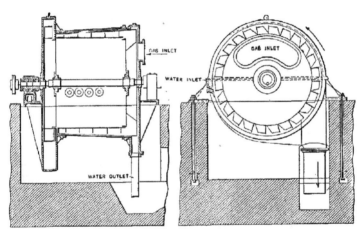

Longitudinal Section.   Cross Section.
The Theisen Centrifugal Gas Washer.

at several furnaces, consists of a long tank with a hopper-shaped section which is kept full of water. The gases are brought down to the surface of the water repeatedly by a series of goose necks, and all dirt, except almost impalpable powder, is removed with a loss in temperature of only 25 degrees F. Iron Jan. p. 6.

For washing gases for use in gas engines the **centrifugal machines** are so far most successful. As they are expensive to operate, they should be used for only that part of the gas which is to be used in the engines. Ten thousand cubic feet per minute is needed to supply engines of 5000 to 6000 H. P. The washing of this quantity of gases in centrifugal machines is said

Iron Age,
Aug. 18, 1904,
p 4 to consume 8000 gallons of water and expend 65 H P. in 24 hours. The Theisen centrifugal machine is largely used in Europe for cleaning gases for engines. It consists of rapidly rotating cylinders, in which the gas enters at the end and water at the periphery. A No. 6 washer is said to clean 30,000 cubic feet of gas per minute at expenditure of 4 H. P. per hour. For use in gas engines the gases should be cooled to 75 degrees F., and should not contain over 0 1 grain dust per cubic foot. In this country the first large installation of blowing engines operated by gas, was at the Lackawanna plant at Buffalo. The system of washing in use there consists of four vertical scrubbers, arranged in series and supplemented by centrifugal machines. The quantity of dust is easily reduced to 0.1 grain per cubic foot.

Hoisting Devices.—The time-honored method of raising stock to the top of the furnace was to hoist it by means of two balanced cage-hoists, operated by a single cable, controlled by the drum of an automatic reversing hoisting engine. The stock was loaded on hand buggies at the stock piles by the bottom-fillers, run on the scales and weighed, then run on the hoist and sent to the top, where the top-fillers dumped the buggies and returned them to the hoist. With the advent of larger furnaces, making enormous outputs and consuming corresponding amounts of stock, the limitations of this system became apparent and automatic charging was substituted.

In **automatic charging** the stock is usually sent to the top in self-dumping skips, running up a steeply inclined skipway. The skips are of much greater capacity than the old hand-buggies, and the filling is consequently more rapid. No top-fillers are needed. The use of skips, however, demands special stock house arrangements. The ore and stone which are stored in a series of bins, with discharge chutes at the bottom, are dumped into larries running beneath the bins, which convey them to the skips. The bins for coke are usually arranged to discharge directly into the skip. The larries are provided with scales, so that the ore or stone can be weighed as drawn from the chutes. In this way a larry with two men can handle as much stock as twenty bottom-fillers.

There are several styles of automatic top filling devices. The original arrangement, the Neeland device, was installed at the Duquesne furnaces of the Carnegie Steel Company when they were built in 1896. It includes large cylindrical buckets, each hav-

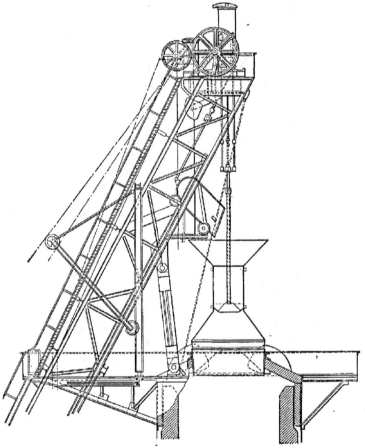

The Julian Kennedy Furnace Top.

ing a small bell for a bottom. The buckets are filled, run up an inclined runway in a carriage, set over the bell and dumped. This method gives excellent results, but has never been duplicated. In its stead the self-dumping skip has been developed. There are several types of the skip hoist in use. They do not differ much in principle. The Brown system operates a single large skip at high speed. The Kennedy system operates two skips on parallel tracks. The Rust system operates two skips, one passing over the other.

The Julian Kennedy furnace top has been adopted by many managers. It has been described and well illustrated by Sahlin. It consists of a double skip-hoist, the skipway being a three-girder structure, carrying double tracks, side by side. These girders are firmly bolted to suitable foundations, and supported at the top by pin-connected struts which bear on the top of the furnace. The skips are of the self-dumping variety, with the bale attached at the rear, and the rear wheels provided with wide treads to engage the outer rails in dumping. Hoisting power may be applied either by the usual vertical type of steam hoist, such as the crane hoist, or by an electric motor-driven drum. Either system must be of the reversing variety and provided with automatic stops to prevent overhoisting. The top of the furnace is made somewhat conical, and is closed by a cast steel lip ring and a Parry bell of the usual type. The whole is then enclosed in a steel plate hood, ending at the top in a cylinder. The bottom of this cylinder is closed by the smaller bell and is surmounted by a rectangular hopper, into which the skip dumps. No explosion doors are provided, as it is claimed that they are unnecessary, and they only serve to allow the escape of stock during slips.

Iron Age, May 14, 1903, p 22.

Stock Distributors.—All skips dump into receiving hoppers which drop the charge on the upper bell. Owing to the fact that in running down an incline lumps tend to outstrip fines, it follows that the side of the hopper farthest from the skip will get coarser material than the near side. Furnaces served by skips often suffer from this cause through the fact that the gas tends to creep up through the more porous portions of the charge and there erode the lining. Every furnace so equipped very soon burns out its lining on one side, developing hot spots on the shell, and thus compelling blowing out and relining. As a result several styles of deflectors have been devised with a view to distributing the stock more evenly.

Iron Age, Jan. 12, 1905.

The **Brown distributor** consists of a hopper in the form of an

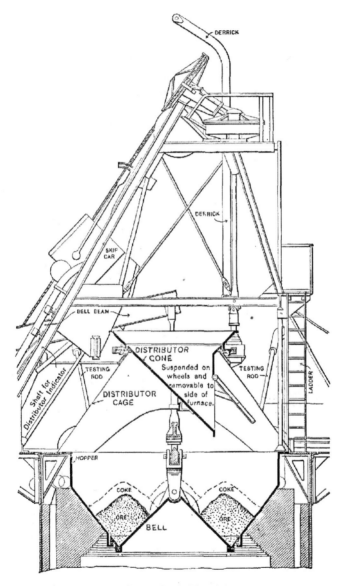

The Brown Stock Distributor.

eccentric chute, which impels the charge to one side of the bell, but which is geared so that it rotates a given part of the circle each time the skip comes up, thus causing the charge to be dumped at successive points in the hopper, as was the case in hand dumping. This device obviates the use of the smaller bell, as a flap automatically closes the distributing chute whenever the bell is lowered

Ibid,
Apr 12, 1900,
p 1

Tr. A. I. M. E,
XXXV., p 569

The **Baker & Neumann** distributing device gets the effect of

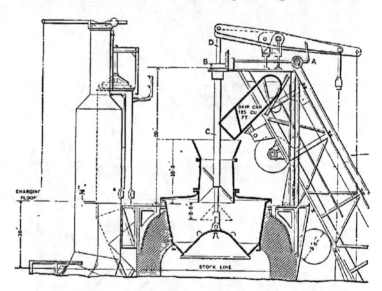

The Baker-Neumann Stock Distributer

the Brown device in a different way. It is provided with a deflector plate, set at an angle and attached to the sleeve hanger which operates the small bell. The deflector is located within the smaller hopper, just above the small bell; and when the bell is lowered, compels all material to fall on one side of the distributing bell. The sleeve carrying the deflector is made to rotate 91 degrees each time the bell closes, thereby depositing the next skipful in a new place. This device has proven its efficiency.

Tr. A. I. M. E,
XXXVII., p. 523

Stock Indicators.—In hand charging a furnace it is the duty of the top-fillers also to gauge the stock occasionally to see that the furnace is kept full. This is done by feeling with a slender iron rod through small holes left for this purpose in the furnace top In the case of automatic charging, however, it is necessary to have some form of automatic device to take the place of the top-filler.

The **Baker Stock Indicator** consists of a steel rod, which passes through the gauge hole to the top of the stock, and is suspended freely over pulleys by means of a flexible connection which leads down to the ground, and indicates the position of the rod. Iron Age, July 18 1901, p. 13.

In the **Johnson Automatic Stock R corder** the flexible connection is passed over a drum which controls a dial on the stockhouse A recording pen makes a permanent record, which is a continual check on the fillers. By means of a rack and pinion the rod is automatically lifted clear of the stock when the bell is lowered Ibid, May 14, 1905, p 1441

HOT BLAST STOVES.

Hot blast stoves are devices for heating the blast. They are usually located at some point between the blowing engine and the furnace in order that the blast may conveniently pass through them on its way to the furnace. They should be placed as near the furnace as practicable, as the waste gases are always used as the fuel for heating them

Formerly all furnaces were blown with cold-blast, and special grades of iron are still made in that way But since the hot blast was first introduced by Neilson in 1828, its use has been practically universal For many years the heating was accomplished by what is known as the " **iron pipe stove.** " It consisted of a series of rows of cast iron U pipes, inverted and set in connecting foot pieces, forming continuous passages of great length, around which the gases were burned. The blast, passing through the pipes, absorbed the heat transmitted to it from the burning gases. The temperature of the blast from such stoves is limited to about 900 degrees F., as the pipes deteriorate rapidly when heated higher. There are still many of them in use at old furnace plants.

The " **firebrick** " **hot blast stove** came into use during the sixties. It has the advantage of being able to attain a temperature of 1,500 degrees F. It operates on the Siemens regenerative principle, and hence is intermittent in its action. It is first heated by burning gas in it, and then used for heating the blast which passes through it in the opposite direction. There must be at least two such stoves to each furnace, so that one may be heating while the other is in use. Such an arrangement, however, would not maintain a very high temperature, since each would be in use half the time. It is necessary, therefore, to have at least three stoves to a furnace, in order that the period of heating may be longer than the period of use. The latest construction always allows for four, in order that the heating capacity may be ample during cleaning and repairs.

The exterior of a stove consists of a tall, cylindrical, riveted steel plate shell with a dome-shaped top. The stoves are usually as high and wide as the furnace itself. The extreme of size to date is that of the stoves of the Lackawanna Company at Buffalo, viz., 135 feet by 22 feet. The shell is pierced for suitable inlet and outlet valves for both gas and blast, as well as air inlets and blow-off valves.

The interior of the stove is of firebrick construction, and is of many different designs. All firebrick stoves consist primarily of two parts, the combustion chamber and the checkerwork. The combustion chamber is an open space reaching to the top of the stove, at the bottom of which the gas is introduced and. burned. The checker work consists of a series of straight parallel firebrick flues leading from the top to the bottom of the stoves. The hot products of combustion from the combustion chamber pass through these flues, giving up their heat to the walls, where it is stored for use in heating the blast. The blast is heated by passing it through the stove in the opposite direction. It is customary to use a stove for an hour for heating the blast, then to allow it to absorb heat until its turn comes again. This may be for two or three hours, according to the number of stoves.

Classification of Stoves.—Stoves may be classified according to the location of the combustion chamber. The **side combustion** chamber consists of a segment of the stove situated at one side and separated from the checkers by a heavy division wall, arched

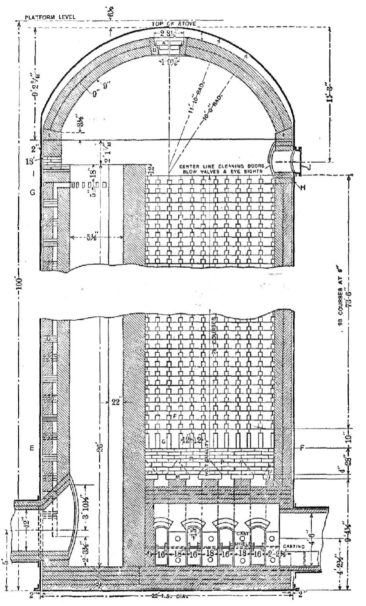

The Roberts Hot Blast Stove.

toward the checkers. The **central combustion** chamber, as the name implies, is situated concentrically in the centre of the stove. The checkers are located all around it.

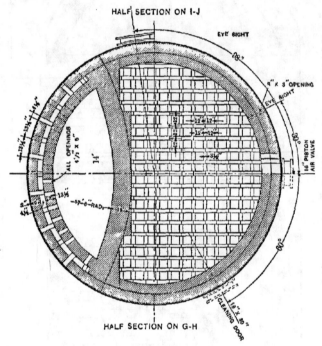

The Roberts Hot Blast Stove.

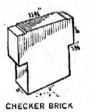

CHECKER BRICK

Stoves may also be classified according to the number of times the products of combustion pass through them. A **two-pass stove** contains a combustion chamber and one-pass of checkers, a

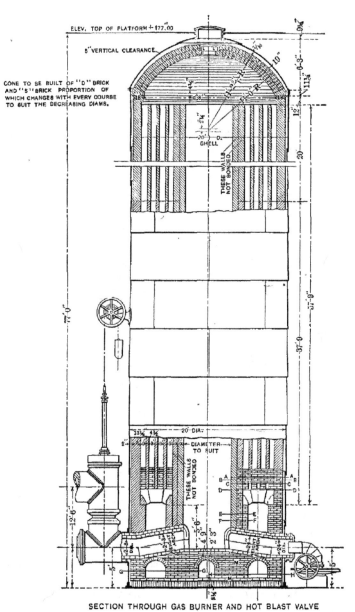

SECTION THROUGH GAS BURNER AND HOT BLAST VALVE

The Julian Kennedy Hot Blast Stove.

three-pass stove has two passes of checkers, and a **four-pass stove** has three passes of checkers.

The first development of the firebrick stove took place in England, in the form of the Whitwell type. The original Whitwell stove had a side combustion chamber and 10 small passes of checkerwork. When introduced into this country, however, the number of passes was reduced to 4. The type has now become

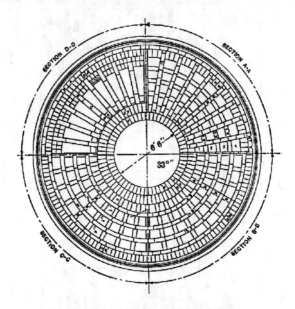

The Julian Kennedy Hot Blast Stove.

practically obsolete. A parallel development was the Cowper stove, which was a two-pass stove with either side or central combustion chamber. This type, under different modifications came to be used almost universally, first with side, and later more generally with central combustion chambers. There is little to choose between them. The side chamber gives a greater checker area, but the central chamber loses less heat by radiation from the combustion chamber. There are several surviving modifications of the Cowper type of stove, among which may be mentioned those advocated by Roberts, Kennedy and Foote. The Roberts stoves are usually side combustion stoves of two passes; the Kennedy stoves, while having two passes, are usually central com-

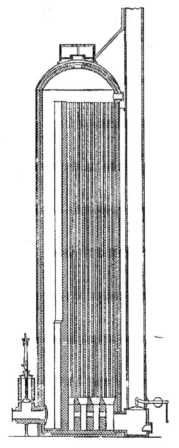

The Foote Hot Blast Stove.

bustion stoves. The Foote stoves also are two-pass, side-combustion chamber stoves, and constructed of patented mitre-shaped bricks.

In all of these stoves, the gas burns up through the combus-

tion chamber, and the products of combustion pass down through
the flues of the checker work to the chimney flue. When the stove

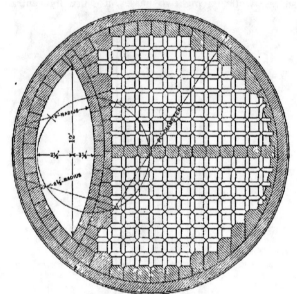

ENLARGED CROSS SECTION
The Foote Hot Blast Stove.

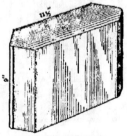

CHECKER BRICK

is in use, the blast is admitted by a valve at the bottom of the
checkers and pursues the reverse course, passing out near the
bottom of the combustion chamber.

The three-pass central combustion stove, used by Massick and

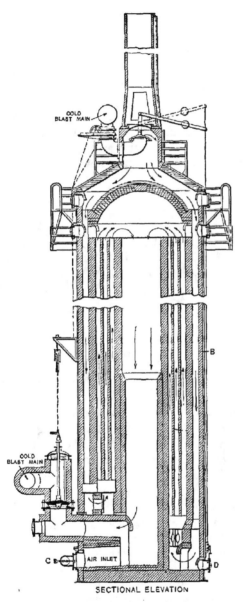

SECTIONAL ELEVATION

The McClure-Amsler Hot Blast Stove.

117

Crookes and modified by McClure and Amsler, has a central chamber similar to that of the two-pass stoves. The products of combustion are deflected by an inner dome and made to pass down an annular set of checkers to the bottom of the stove, only to pass upward through an outer annular set of flues which discharge into an individual chimney surmounting each stove. The blast must, of course, be admitted at the top of the stove and pursue the opposite course. It is claimed that this arrangement cools the gases more thoroughly than the two-pass stoves, and the individual,

Iron Age, May 5, 1892, p. 864.

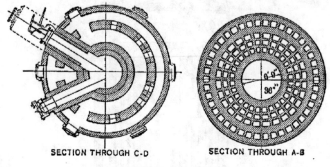

SECTION THROUGH C-D   SECTION THROUGH A-B

The McClure-Amsler Stove.

symmetrically-placed chimney permits a more even heating of the flues.

Stove Chimneys.—The hot blast stoves run usually in a straight line from the furnace toward the blowing engine house. The draught chimney may be at the end of the row, but more often is midway. In some instances individual chimneys are used for two-pass stoves, because when a single chimney is used, the stove nearest it gets the best draught to the detriment of the others. Neither style of chimney is entirely satisfactory in the case of central combustion stoves, as the pull of the draught tends to utilize the checkers nearest the chimney flue to the exclusion of the others, and thereby reduces the working area of the stove. This objection is also apparent in side combustion stoves. It is entirely obviated in three-pass central combustion stoves by having a central chimney on top of the stove, thereby giving a symmetrical draught.

Stove Valves.—The valve construction of hot blast stoves presents a great variety of forms There are four valves to be considered, viz., the gas valve, the chimney valve, the cold blast valve, and the hot blast valve The **cold blast valve** is subjected to the least trying conditions. Since it simply admits air at temperatures somewhat above that of the atmosphere, a simple gate valve is all that is necessary. The **gas valve** is exposed to somewhat higher, but not excessive temperatures. They should never exceed that of the furnace top, from 500 degrees to 600 degrees F. This valve and burner are usually the Spearman type. The temperature to which **the chimney valve** is subjected depends upon the degree to which the checker work cools the products of combustion. This ranges from 400 to 1000 degrees F. in different stoves. This valve is generally a water-cooled, mushroom valve, although a simple adaptation of the Spearman gas valve is meeting much favor. The **hot blast valve** is subjected to the most trying condition of the four. When the stove is on blast, this valve is constantly in the presence of air at a temperature far above visible redness, and is, therefore, the most vulnerable point in the system. It is generally of the mushroom type, water-cooled and having a water-cooled seat. A mushroom valve is usually a hollow iron casting, although sometimes it is made of bronze or other metallic composition. It has a hollow stem of the same metal, which is bolted to or screwed into it, and by which it is raised or lowered. Through the stem passes a pipe for admitting the cooling water, which finds egress through the annular space between the pipe and the stem. The valve has a circular bearing turned to fit the valve seat. This, being a hollow casting, is also water cooled. The bearing of the valve on the seat should be accurately adjusted, as any leakage of the blast permits rapid wear of the metal, especially if the gas contains much dirt. The Berg hot blast seat, which is widely used, consists of a hollow water-cooled ring, clamped between two steel castings which face the brick work. By loosening the clamps it may be readily removed and replaced.

Iron Tr
Review,
Feb 1, 1

The most usual style of **gas burner** is some modification of the Spearman type It consists of a movable gooseneck, rising out of an underground gas flue, and leading horizontally into

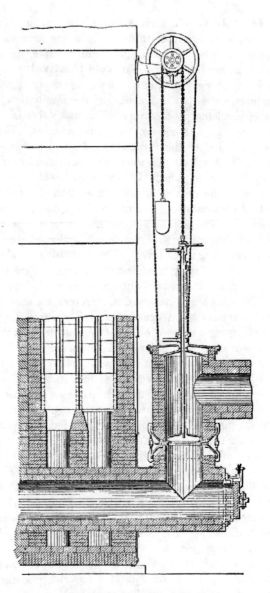

The Kennedy-Morrison Hot Blast Valve.

a gas port in the side of the stove. When not in use, the goose-
neck is racked back and the port closed. When in use the port
is opened, and the gooseneck racked forward, automatically
opening the outlet of the gas flue. This device is cumbersome,
but it effectually prevents the leakage of gas into the blast with
the attendant danger of explosions. The Kennedy-Morrison
chimney valve, which is being widely applied, is of the same
general form as the Spearman burner.

Regulating Valves.—As a rule, the hot blast main is on the
same side of the stove as the gas main, and the cold blast pipe
on the same side as the stack flue. It is usual to arrange a by-
pass pipe to connect the cold blast and the hot blast mains.
This pipe is fitted with a controlling valve, by means of which
cold blast may be mingled with the hot blast to modify its temperature when desired. A valve and escape pipe, known as the
" snort valve," which can be controlled from the cast house, is
always placed in the cold blast main. It is designed to permit
the throwing off of the blast for any temporary cause, such as
closing the tapping hole, without stopping the engines. Usually
a butterfly check valve is arranged to close automatically when
the snort valve is open, in order to prevent any gas backing up
to the blowing engines. This is also the object of a gas escape
valve, usually placed in the hot blast main near the bustle pipe,
which drops automatically when the pressure falls.

Equalizers.—When a fresh stove is put on blast, its temperature may be two or three hundred degrees above that of the previous stove when taken off. The temperature of the new stove
will drop gradually during the hour in use. At the next change
there follows another sudden jump in temperature. Such
changes add to the irregular conditions under which furnaces
operate. To obviate them, the idea of an equalizer was evolved
in England in 1901 by Gjers and Harrison. Their equalizer con- _{Inst. Jour.,}
sists of a steel plate shell, filled with checkerwork, and not unlike _{1902, II.,}
a stove in appearance. It is 55 feet high and 20 feet in diameter. _{p. 282}
The checkerwork is divided vertically in the middle by an impervious wall, reaching nearly to the top, thus giving a two-pass
arrangement. The equalizer is placed between the stoves and
the furnace, and the heated blast passes through it on its way to ·

the furnace. The high temperature of the blast from a fresh stove is stored up to be given out later when the stove temperature has dropped. In this way an almost unvarying blast tem-

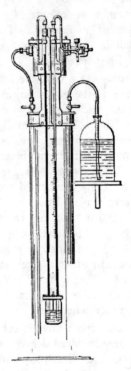

The Single Uehling and Steimbart Pyrometer.

Iron Age, Aug. 18, 1904, p. 6. perature is obtained. A small equalizer, 20 feet $\times$ 12 feet, has recently been installed at Stanhope, N. J.

PYROMETRY.

Pyrometry.—Furnaces are usually provided with a pyrometer for measuring the temperature of the blast and escaping gas. Of the many varieties in existence the **Uehling and Steinbart** is most generally used in this country. It depends for its principle of action upon the difference of volume of air when heated and when subsequently cooled. A chamber having an inlet opening and a

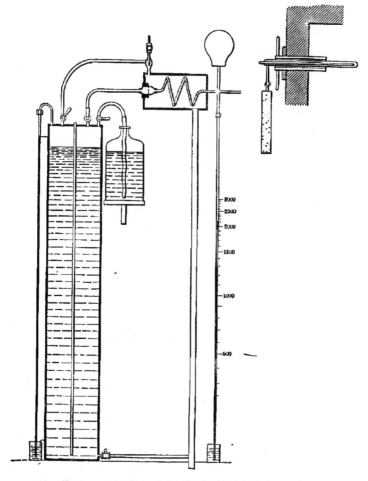

Diagrammatic View of the Uehling-Steinbart Pyrometer

manometer tube dipping into water, connects with another chamber having the same sized inlet and a similar manometer tube. The latter chamber is subjected to uniform suction. The heated air is

Iron Age,
Feb. 22, 1894.
drawn into the first chamber by the suction of the second, but it is cooled to 212° F. before reaching the second chamber. The partial vacuum thus created is indicated by the difference in reading of the manometer tubes, which serves as a measure of the difference of temperature of the blast before and after cooling to 212 degrees.
Inst. Jour.,
1904, I.,
p. 124.
The tube which receives the heated air is usually tapped into the hot blast main before it reaches the bustle pipe, but the manometers may be stationed at any convenient point. By means of a pen-point a continuous record of temperatures may be kept which serves as a check on the hot blast conditions.

Another type of pyrometer is the **Le Chatelier.** It depends for its action upon the fact that when a pure platinum wire is

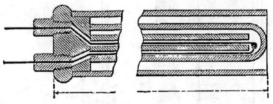

Le Chatelier Pyrometer.

fused to one alloyed with 10 per cent. rhodium or iridium, and the joint is heated, an electromotive force is set up whose strength is in proportion to the temperature which creates it. If these two
Ibid,
p. 106.
wires are connected to a galvanometer stationed at any suitable point, the temperatures may be determined by means of the galvanometer deflections. Such a thermo-electric couple should be encased in a porcelain tube for protection before being subjected to high temperatures. It has the advantage of not being impaired in efficiency by length of connections. The rhodium alloy is said to be the more durable of the two, but the iridium gives greater deflection to the galvanometer needle.

The **Brown pyrometer** was for a long time used almost exclusively. It depended for its action upon the expansive force of heated metal. A thin strip of iron, placed in a tube, was fixed firmly at one end and attached to a dial hand at the other. The hot blast passing through the tube, expanded the iron strip, and the amount of expansion was indicated by the dial hand.

THE CAST HOUSE.

The cast house is the building which shelters the pig-bed. It was formerly built of bricks, with arched doorways, but is now generally of steel construction, and is often provided with overhead travelling cranes for handling the iron and for other pur-

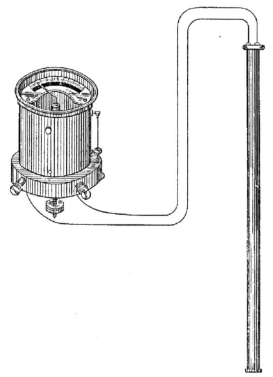

Le Chateller Pyrometer and Galvanometer.

poses. Except in the case of furnaces making foundry irons exclusively, the use of ladles and pig-casting machines for handling the product has obviated the necessity for the cast house. As a result, it has degenerated, in the latest plants, into a mere covering for the runners and spouts which convey the metal to the various ladles.

THE BOILER PLANT.

The boiler plant of a blast furnace furnishes the power for blowing the blast, for pumping the water, for hoisting the stock, and also for furnishing the electricity for power or lighting. The fuel is the blast furnace gases, sometimes supplemented by coal. As no gas can be produced without power, it is necessary, when blowing in a furnace, to start the boilers on coal. For this reason the boiler plant is always provided with coal grates as well as gas burners. Any standard type of boiler is suitable for a blast furnace plant. The **Babcock and Wilcox,** the **Sterling,** and the **Cahall** are perhaps those most universally used, although many others also give excellent satisfaction. A furnace usually requires about 8 boiler horsepower for each ton of coke burned in 24 hours.

BLOWING ENGINES

Blowing engines are used to forced the blast into the furnace. Until recently they were usually driven by steam, but during the past few years very important gas driven installations have been made. The turbo-blower is also being introduced.

Blowing engines are of three general types: vertical, horizontal and vertical-horizontal.

There are three usual modifications of the **vertical type.** The simplest form consists of a simple engine having its air cylinder directly above the steam cylinder with both pistons on the same rod. It is provided with a fly-wheel on each side, whose crank rods are connected with the piston rod by means of a long cross-head, on account of which it is generally known as the " long crosshead " engine. The steam end may be made either high or low pressure, and such a pair may be operated as a compound engine.

The most common modification of the vertical type at present is the vertical cross compound connected engine, which consists essentially of a pair of vertical engines provided with a single fly-wheel between them, which obviates the use of a long cross-head. They have the air cylinders above the steam cylinders, as in the long crosshead engine, and owing to their great height are commonly called the " steeple " engines.

The Vertical Type Reynolds Compound Engine,

The Vertical Disconnected Type, Southwark Quarter Crank Blowing Engine, Compound Disconnected.

The Horizontal Type, Mesta Compound Engine.

The third modification of the vertical type is a simple engine with its steam and air cylinders on separate pedestals. Consequently the respective pistons are operated by separate rods, which connect with a shaft bearing a single fly-wheel The two cranks are usually set 90 degrees apart so that the maximum effort coincides with the maximum resistance, thereby promoting smoothness of action. They are, therefore, generally known as the " quarter-crank " engine. As high and low pressure engines may be operated in pairs, they are more definitely described as vertical cross compound disconnected engines.

Engines of the **horizontal** type may also be built single, compound connected, or compound disconnected They are generally built compound connected, having the steam and air cylinders placed tandem with their pistons operated by a single rod driving a single flywheel between, with cranks set 90 degrees apart, quite similar to the vertical compound connected engine

The **vertical-horizontal** type of engine may also be simple or compound. Usually the air cylinders are placed vertically and the steam cylinders horizontally in the same vertical plane. Since the piston rods are independent of each other their cranks may be set 90 degrees apart, and so be classed also as quarter-crank engines.

The principal makers supply engines in all these different types They are designed to deliver 40,000 to 60,000 cubic feet of air per minute They are usually compound condensing engines having high pressure cylinders, 40 to 48 inches in diameter, and low pressure cylinders 78 to 84 inches in diameter. The blowing cylinders are usually 84 to 96 inches in diameter, and the stroke is 60 or 66 inches To equalize the motion they have heavy fly-wheels which are designed to run as high as 50 revolutions, against a maximum pressure of 30 pounds per square inch, on 125 to 150 pounds steam pressure. The use of Corliss steam valves is practically universal. The chief feature which distinguishes different makes of engines is the type of air valve.

The **Koerting gas engine** is used by the Lackawanna Steel Company, at Buffalo. It is a double-acting two-cycle engine of the horizontal-vertical type, having a horizontal gas cylinder and a Southwark vertical air cylinder, and provided with a heavy fly-

The Vertical Horizontal Type Tod, 50 x 90 x 90 x 60 Cross Compound.

wheel. The length of the gas cylinder is about double the stroke of the engine, and the length of the piston is about half that of the cylinder. Each end of the cylinder is provided with a special sloping head, carrying inlet valves. The exhaust takes place through a series of open ports midway in the cylinder The length of the piston is so regulated that the exhaust ports are not uncovered except at the extreme end of the strokes Two auxiliary cylinders, acting as simple piston valves, are arranged to pump the gas and air respectively to both ends of the working cylinder in quantities suitable for mixture. They are driven by a crank 110 degrees in advance of the main crank. The inlet valve is opened before the main crank reaches the dead center. The air enters first and part of it goes to sweep out the exhaust gases and form a cushion between them and the fresh gas. The new mixture is compressed to 10 atmospheres by the returning piston and at dead center two electric igniters start combustion. The engines are rated at 2000 horsepower each, and deliver about 20,000 cubic feet of air up to 30 pounds pressure per minute of 60 revolutions.

CHAPTER III.

OPERATION OF THE FURNACE.

Introductory.—In no business are good judgment and long experience more important than in the operation of a blast furnace. They are necessary, not only to the manager, but to each employee. The organization of a blast furnace force must be such that each person has his specific duties, for which he alone is responsible. Each turn is in the care of a general foreman or **founder,** sometimes called the "blower," who maintains a general supervision over the furnace and all of its accessories. The furnace itself is in charge of a "**keeper,**" who supervises the tapping of iron and cinder, and also watches the condition of the furnace both within and without   He is usually provided with two or more "**helpers,**" who assist about the furnace. The skimmer and runners are the especial care of the first helper, the pig bed and the running of the iron of the others   The care and disposal of the cinder and slag are in charge of **slag men,** or "cinder snappers," as they are generally called   A **stove tender** looks after the hot blast stoves, and the stock is handled by the **fillers.**

BLOWING IN.

Drying the Furnace.—When a newly-built furnace, or an old one which has been freshly lined with bricks, is about to be put into blast, it should first be subjected to careful drying to remove all moisture from the new brickwork. This is usually accomplished by means of a wood fire built in the hearth. The drying should be gradual to prevent any undue shrinkage or cracking of the brickwork. It is begun by means of a light fire, which is gradually increased to sufficient intensity to insure completeness. Formerly it was thought necessary to consume two weeks in the operation, but now many managers consider one week sufficient

Filling the Furnace —The operation of smelting iron is inaugurated by filling the furnace with cold stock and subsequently lighting it. As failure in properly lighting a furnace may entail great expense, the filling should be done carefully. The general method consists of first placing wood on the hearth, and following it with a large body of coke, upon which is built a succession of light but gradually increasing charges of ore, properly fluxed. The first layer of wood is sometimes deposited directly on the hearth, but usually on a scaffold about the level of the tuyeres. It consists of cordwood sticks placed on end. Sometimes as many ' as three such layers of wood are used, thus bringing the top of the wood well up into the boshes. Upon the wood is placed a blank charge of coke, mingled with a sufficient quantity of limestone to flux its ash, and 10 to 25 per cent. of gray blast furnace cinder to prevent too high alumina in the slag. This blank charge of coke varies usually from one-third to one-half the cubical contents of the furnace. It is advisable to charge plenty of coke at this point so that the furnace may be hot and gray from the start. Upon this bed of coke, the charges of fuel, ore and flux are begun. The first five to ten rounds of fuel are accompanied by light charges of ore, usually about half the weight of the fuel. Sufficient limestone to flux the gangue of the ore and the ash of the fuel is added, together with an equal quantity of blast furnace cinder. With each subsequent series of five to ten rounds of fuel, the ore and stone are increased, and the cinder diminished, until by the time the furnace is full, the ratio of fuel to ore is about 1 : 1. After blowing in, the ratio is gradually increased from day to day, as fast as is warranted by the temperature of the hearth, as shown by the quality of cinder and iron produced. Usually the normal burden is reached within a week or ten days.

Lighting the Furnace. –When the furnace has been completely filled it is ready to light. The space beneath the scaffold is filled with kindling, and a quantity of kerosene oil is poured into each tuyere. The bell is closed and the bleeder opened, and all the gas burners at both boilers and stoves are closed tight. Light blast is then turned on, and fire is started simultaneously at all the tuyeres by means of a red-hot iron rod thrust through the pricker hole. In two or three minues, smoky gas appears at the top of the furnace. This should be ignited at once. The bell is kept closed and the bleeder kept open until all of the wood smoke disap-.

pears and the gases burn clearly and freely at the top. Then the bell is closed and the gases find their way into the downtake The gas burner farthest from the furnace on the boiler line may then be cautiously opened till the gas ignites As the volume of gas increases, other burners may be turned on. These precautions are imperative in order to avoid the disastrous explosions which frequently result from igniting gas before all the air has been swept out of the pipes   The gases are especially dangerous during blowing-in. Explosions are extremely liable to happen and the effects of inhalation are rapid, violent and sometimes fatal. These phenomena are due to the abnormally high percentage of CO in the gases while the coke blank is burning and reduction is not rapid. Analyses of gases which are given off by the furnace during blowing-in show that when the furnace is first lighted. combustion is nearly complete, and little CO is formed. After an hour or so, the mass of coke becomes heated to-the temperature where it can react upon CO_2, reducing it to CO. The ratio of CO to CO_2 may then rise to 10, making the gas very dangerous. The ratio soon falls to about 5, whence it is gradually reduced as the burden rises to normal.

Danger in Carbonic Oxide.—The effect of CO on the human system is highly poisonous. Even when very much diluted by air it is dangerous, and its effects are cumulative. It is said that exposure for an hour to air containing small quantities of it produces the following result:

> 0.20 per cent. produces giddiness.
> 0.35 per cent produces inability to walk.
> 0 70 per cent produces unconsciousness
> 1.00 per cent. is very dangerous.

When in its natural state, blast furnace gas is both visible and odorous. When cooled and washed it is neither, and hence is very insidious, particularly when used in gas engines. It permeates soil and masonry, especially when heated. Hence they should be rendered impervious by cement or concrete. Leaks may be detected by introducing in the mains a strongly odorous substance, such as acetylene. Mice and birds are quickly affected and so may serve as tests for doubtful places.

Tapping the Furnace.—Within 10 to 15 hours after the blast is put on, the slag will have accumulated in sufficient quantity to reach the level of the cinder notch. By watching through the peep hole, the keeper can see when the slag rises to the level of the tuyeres. The cinder is then flushed. Three or four flushes usually take place before much iron accumulates in the hearth. Usually 20 to 30 hours after the lighting of the furnace, the iron is ready to tap. The tap hole is generally opened by means of a hand drill, but it is being replaced in some localities by a motor drill operated by compressed air. When the drill has cut away the clay until it shows bright red, a bar is driven into the hole with sledges. The opening so made is quickly enlarged by flowing metal, which is then directed by the helpers into its proper channels.

METHODS OF HANDLING PRODUCTS.

There are two general methods of handling the iron which flows from the furnace, viz.: cold, in pig-beds, and molten, in ladles.

CASTING IN PIG-BEDS

Sand Pig-Beds.—Formerly all the iron was handled in sand pig-beds. By means of wooden patterns a series of parallel, adjacent depressions, about 40 inches long, 4 inches wide and 4 inches deep, are moulded in loose, moist sand. These depressions are connected to the main runner, which leads from the tap-hole by means of a cross runner which connects with one end of each depression. Into these depressions the molten iron is led. Owing to a fancied similarity in appearance, the iron in the cross runner is known as the " sow," and that in the parallel depressions as the " pigs." As soon as the iron is fairly set, it is covered with a layer of sand which is scattered over it by the shovelful. Then, by means of bars and sledges the pigs are broken from the sow, and the sow broken into convenient lengths. The purpose of the sand covering is twofold: to protect the workmen from the intense heat of the iron, and to retard the cooling, thereby facilitating breaking the iron and incidentally increasing the size of the grain. The pigs are then cooled by a spray of water, loaded on trucks by hand and taken to the wharf where they are broken and piled, ready for shipment. This method of handling the iron permits cooling in a bed of slowly-conducting material, thereby allowing large crystals to develop in the pig. On the other hand, it permits the adhesion of much sand, which is undesirable in iron intended for basic steel manufacture The sand pig-bed is destroyed by use, and must be remade before each cast.

Chills.—A modification of the sand pig-bed is the iron pig-bed, known as " chills." It comprises a permanent pig-bed, which is composed of heavy iron castings moulded in shape very similar to that of the sand pig-bed. Such a bed requires no preparation beyond sweeping and sprinkling with clay wash before each cast. The heat of the chill dries the wash, leaving a coating of clay, which prevents sticking and cutting or melting of the chills. The iron is run into it, cooled and broken in the same way as in the sand bed. The resulting iron is free from adhering sand; hence this system is advantageous in making iron for steel manufacture.

Pig Breakers.—Sometimes pigs are handled by overhead cranes, which transport the unbroken sections of the cast to a pig-breaker. Such a device was installed at the Duquesne plant, and is still occasionally used. The pig-bed, instead of being in the usual form, consisted of long, parallel pigs or sows, about 20 feet long, joined together in pairs. This was a convenient shape for the breaker. The breaker consists of a heavy plunger, working over a table by means of an eccentric shaft. The breaker is fed by a bed of motor-driven live rolls, that bring the piece to be broken under the oscillating plunger. The broken pigs slide from the breaker into cars.

At the plant of the Buffalo Susquehanna Company the iron is run in sand beds and when cooled sufficiently is picked up by a travelling crane and taken to a Brown Pig Breaker, where it is broken for shipment.

CASTING IN LADLES.

As furnaces increased in size and capacity, they produced such enormous quantities of iron at each cast, that the pig-bed system became impracticable. Moreover it was found to be cheaper in the case of steel manufacture, to take the iron directly to the steel furnaces in the molten condition. For these reasons, the system of handling it in ladles was adopted. This method has come into

such general use that at present a large percentage of all pig
intended for steel making and also some foundry pig is tapped
into ladles, and a decreasing percentage of the total output of
the country is run into pig-beds.

For the use of the ladle system of handling iron, it is necessary
that the iron runner should branch to a series of spouts at least
equal in number to the ladles needed to hold a cast. These spouts
should be at such an elevation that the ladles may be stationed on

The Berg Hot Metal Ladle.

tracks beneath. Each ladle should hold 20 tons or more and still
have margin to allow for slopping. When the ladles are filled the
iron is covered with coke dust and sent to the steel mill.

Hot Metal Ladles.—There are many makes of hot metal
ladles, which differ only in minor details. Those manufactured by
the Pollock Company, of Youngstown, Ohio, from the Berg pat-
ents, will serve as a type. The ladle bowl is made of heavy, riveted
plates, set in a steel cast trunnion-ring, and lined with fire bricks.
The running gear consists of two single trucks connected by a
steel cast frame to which a steel cast buffer is strongly riveted.

Couplers, journal-boxes, etc., are of the usual M. C. B. patterns. The trunnions are in the form of pinions and work in racks, so that as the ladle tilts, it travels forward. The tilting is accomplished by means of a worm gear, which may be operated by hand or by power.

The Treadwell hot metal ladle is constructed on similar lines. Ibid, July 10,

PIG CASTING MACHINES.

If the steel mill were in a position to receive the product of the blast furnace continuously, a series of ladles would be the only

The Berg Cinder Ladle.

handling equipment necessary. But in order to handle the output of iron when the mill is shut down or over Sundays, or in the case of iron that must be allowed to cool for shipment, a further device is needed. For this reason, mechanical pig-beds, known as pig casting machines, have been devised. They were first used by furnaces operated in conjunction with steel plants, but were later adopted by isolated furnaces, even those making foundry pig.

A pig casting machine consists essentially of a series of moulds, which are made to pass successively under a spout into which the molten iron in the ladle is poured. There are several kinds of pig casting machines, which may be classed under two types, the endless chain and the circular disc.

Endless Chain Machines.—The leading examples of the endless chain type are the Heyl and Patterson and the Uehling machines.

The **Heyl and Patterson** machine consists essentially of a pair of endless chains, supported by wheels which run on a track, and which carry a series of pressed steel moulds. The trackway is

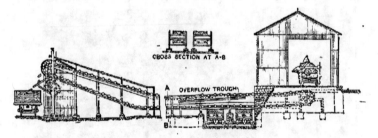

The Heyl and Patterson Pig Casting Machine.

Iron Age,
May 18, 1899,
p. 1.
approximately horizontal, except at the delivering end, where it rises so that the pigs may fall into railway cars as the moulds reverse themselves. A portion of the horizontal track is depressed and passes through a tank full of running water, which cools the pigs enough for handling. After the moulds dump their burden at the delivering end, they return to the receiving end on a lower trackway in an inverted position, passing over smoke ovens or tar swabs, which give them a carbonaceous coating to prevent the iron from sticking. The endless chain is driven by a pinion shaft, each pinion being equipped with a friction clutch. A pair of chains will handle 1500 tons in 24 hours with expenditure of 14 horsepower per hour.

The **Uehling Pig Casting** machine was the first which proved successful. It consisted of a series of moulds on an endless chain which dumped into a tank of water at the turn. The present form is divided into two sections, the first being the casting machine, and the second the cooling conveyor. Each section consists of an independent endless chain. On the casting section the moulds are of cast iron and no cooling water is used. By the time the pigs reach the end of the casting section, they are sufficiently solidified to be discharged to the cooling conveyor. On the return strand the moulds pass over a lime vat in the inverted position and receive a spray of milk of lime, which forms a protective coating. The cooling conveyor is of similar construction, except that the moulds are replaced by steel plates which receive the pigs, carry them through a large trough filled with water and dump them on cars. This section may be set in line with the casting section, or, if economy of space is desirable, at right angles to it.

Both of these machines are now made by the Heyl and Patterson company. The essential differences between them are as follows: The Heyl and Patterson machines use pressed steel pans with a tar coating, and the pig is plunged in water very early. The Uehling machine uses cast-iron pans, coated with lime, and the cooling by water is later. The first cost of the Heyl and Patterson machine is less than that of the Uehling, but the operating expenses and the losses of metal are said to be greater. The cost of maintaining and operating the machines usually ranges from 15 to 22 cents per ton of pig made. This is about equally divided between operating and maintenance.

Disc Machines.—The disc machines have not met with as much favor as the endless chain machines. The leading example is the Davies machine. It consists of a horizontal revolving wheel, about 40 feet in diameter, with a series of moulds on its periphery. The moulds are rectangular cast iron blocks, having depressions on all four sides, which serve as moulds in turn and thereby prolong the service of each mould. The moulds pass under the pouring spout, and the pigs become solidified by the time half the circle is traversed, when they are dumped into a tank of water. The moulds are then sprayed with milk of lime, and are ready for a fresh charge by the time they reach the spout.

SAMPLE FOR ANALYSIS

It is customary to take one or more samples of the iron while it flows from the furnace for purposes of analysis. Proper sampling is very important, in order that analysis may represent a fair average. The duty of sampling usually devolves upon the stove-tender. It is accomplished by means of a wrought iron or soft steel spoon, 6 or 8 inches in diameter, with a 4-foot handle. The spoon is first washed with clay suspended in water, to protect the bowl, and to prevent the metal from sticking to it. It is then dipped cautiously into the flowing metal, and its contents poured into a small mould, making an ingot about 5 x 2 x 1 inch, or poured into a pail of water which granulates the metal in the form of shot. Shot samples are generally slightly lower in Si and notably lower in S than ingot samples. If only one sample is taken it is imperative to wait until the iron has flowed some minutes, in order to insure a fair average, as the composition at the beginning of the flow may differ considerably from that at the end. If several samples are taken, they are distributed through the flow. One should be taken every 5 to 10 tons of metal. The advisability of these precautions becomes apparent from the following analyses:

Am. Soc Test
Materials,
V., p 216,

Ibid,
III., 182.

	Si	S.
1st bed	1 99	0.024
3d bed..	2 17	0.022
5th bed.	2.29	0 024
7th bed	2 46	0 024
9th bed	2.57	0 027
11th bed	2 60	0 023
13th bed	2 66	0 022
15th bed	2.85	0 020
17th bed.	2 36	0 019
19th bed	1 80	0 023

SKIMMING THE IRON.

Since the tapping hole is approximately level with the bottom of the crucible, it follows naturally that what flows from the furnace first is iron practically free from slag. Later, and especially toward the end of the cast, the slag comes freely. Since it is about one-third of the specific gravity of the iron, it floats upon it, and can be easily separated by " skimming." Skimming devices all include a depression in the iron runner, followed by a dam, over which the iron must rise before it can flow to its receptacles. The skimmer is suspended across the runner over the depression at such a height that it rests on the top of the stream of iron and effectually prevents the slag from being carried over the dam.

The slag is allowed to overflow at the side, and is conducted to suitable receptacles.

Formerly it was customary to mould a dam in a sand runner before each cast, and an iron plate skimmer was suspended across the trough and was raised or lowered during the cast as occasion demanded. The iron left in the trough after casting was drained off by tearing down the dam with a hook. This iron, mixed with

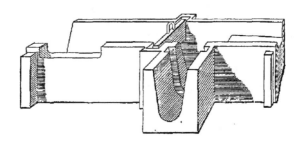

General View.

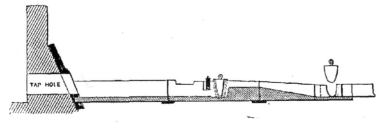

Section of Skimmer.

The Killeen Skimmer and Metal Trough.

sand, flowed into the beds or ladles. The making of a dam was always a delicate task, and a test of a furnaceman's skill. A bad dam might result in considerable loss, by allowing the iron and slag to mix, unless the skimmer was manipulated quickly. The constant watching of a skimmer is not a pleasant occupation, particularly when lead or arsenical ores are used. While furnaces remained small and comparatively small quantities of iron were tapped every six hours, this method was adequate. With rapidly growing furnaces and outputs came changes in skimming devices.

Killeen Skimmer.—The Killeen skimmer was adopted by the Carnegie Company, and the use of its various modifications is fast becoming universal. It consists of a permanent cast iron trough, made in the shape of the old sand runner and dam. Just in front of the dam are slots cast in the side of the trough to receive the skimmer. A depression in one side of the trough in front of the skimmer allows for the overflow of the skimmed slag. An outlet in the side of the trough, between the skimmer and the dam is provided for draining the iron after the cast. To prepare such a dam no skill is needed beyond the ability to smear it with clay to prevent corrosion. The trough is deep and wide, and is amply able to care for sudden rushes of iron or slag. After the iron has ceased to flow from the furnace, the drain gate is raised slightly to allow the iron in the trough in front of the dam to drain into the ladles.

CINDER OR SLAG

Blast furnace cinder and slag are identical in nature. They comprise the molten, non-metallic products of the operation. In practice, however, there is a distinction made between them which has no foundation in difference. What is generally known as "cinder" is flushed from the cinder-notch between casts. The cinder which accompanies the iron at casting time, however, is distinguished by the term "slag." Properly speaking, it is all slag, since slag is the scoria from any smelting operation. The term "cinder" probably resulted from the fact that it includes the ashes and cinders of the fuel which must be removed by-fluxing, since they cannot be raked out as in the case of ordinary fires.

Disposal of Slag.—Three general methods for the disposal of slag may be distinguished; in gutters, in ladles, and in granulation pits.

Allowing the cinder to cool in depressions in the earth is an old and crude method of disposal. It is expensive, as it entails much labor in subsequent loading for removal. But it is sometimes adopted when the slag is to be used for filling and grading purposes, as it cools in small masses which are easily broken. As a regular practice this method was long ago replaced by the use of cinder cars, some of which are still used. The cars had a cast iron body of small capacity into which the cinder was run, and allowed to cool. The cakes of cold cinder were then hauled to the dump.

The almost universal practice at present is to catch the molten cinder in large ladles, run them out to the cinder bank and empty them before solidification takes place. For this purpose the Weimer cinder ladle has long been the standard form. It consists of a large plate steel riveted pot, securely bolted in a cast steel trunnion ring, and lined with brick, or a cast iron thimble made in one piece and having 200 cubic feet capacity. The trunnion is supported on two double trucks of the usual railroad pattern, which are connected by a steel cast frame, securely riveted to buffers, and having standard railroad couplers and journal boxes. It may be tipped to either side by means of worm gear operated by hand or power.

The Pollock cinder ladle is constructed on similar lines. Iron Age, Aug 14, 1902,

The Hartman end-dump cinder car has been less generally used than the side-dump type. It found considerable favor be- Ibid, Dec 27, 1894, p. 1157. cause it could be used to build a cinder bank forward while the other cars poured to the sides. It has a semi-cylindrical bowl which is lined with firebrick, and is tilted forward by means of a pole attached to the locomotive.

Of late there has been a decided tendency toward the granulation of cinder. When slag is allowed to run into a brick or cement pit, and a strong, flat stream of water is made to strike the stream of slag from behind as it falls, the slag is chilled immediately and falls into the pit in a granular condition. It may then be scooped out by means of a clam-shell or orange-peel bucket. loaded on cars and used for filling or road material. When granulated by water at a pressure of 80 to 100 pounds per square inch, slag makes an excellent sand for mortar and concrete. Granulation increases the volume of cinder to three or four times that when molten, and hence requires greater carrying capacity. It also causes a great volume of steam in the cast house during casting, which is a decided disadvantage.

The relative costs of handling cinder under the three systems are roughly as follows:

In gutters, 25 to 30 cents per ton of pig.
In ladles, 12 to 15 cents per ton of pig.
By granulation, 5 to 6 cents per ton of pig.
Two-thirds ton cinder to a ton of pig

The quantity of cinder and slag produced by a blast furnace is generally about one ton for every two of iron made. Many attempts have been made to utilize this vast quantity of material, but even yet the greater part of it goes over the dump. It is said that in some countries bricks can be made of slag as it comes from the furnace, but the tendency of good cinder to slake renders that use doubtful in the United States. Cold cinder which has been broken makes excellent railroad ballast or filling, and is a suitable substitute for crushed rock in making concrete or roofing. Granulated slag mixed with 50 per cent. slaked lime, moulded and treated with steam at 100 pounds pressure makes satisfactory bricks. Excellent cement is made by mixing crushed granulated cinder with 35 to 40 per cent. lime.

News,
p. 384

Care of the Notches.—When the iron flows from the furnace evenly and not too rapidly, it is usual to allow the blast to continue, or at most to slacken it only somewhat. If the iron comes more rapidly than it can be handled to advantage, the blast must be turned off entirely, to lessen the pressure on the molten materials. When the flow is nearly completed, the blast is put on again, in order that the additional pressure on the surface of the fluid in the hearth will compel more of it to flow from the furnace. When the furnace is drained as much as is practicable, the tapping hole is closed.

There are two general methods of closing the tapping hole. The original and until recent years the universal method was by means of balls of wet clay or clay mixed with 10 to 20 per cent. coal or coke dust, thrown into the opening by a helper and rammed back by a **stopping hook** in the hands of the keeper. It was necessary to throw off the blast completely before this task could be attempted. The later and more improved method consists of using a large pneumatic or steam gun to shoot clay into the opening. With the gun it is usually not necessary to cut off the blast completely. The **Vaughn gun,** which is universally used, consists of two cast iron cylinders, which are connected by a cast iron distance piece, and whose pistons are joined by a single piston rod. The steam, acting upon the piston in the steam cylinder, forces

Iron.
Nov.

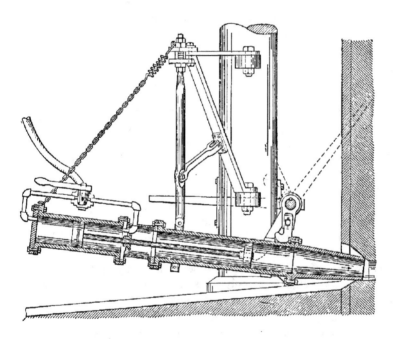

The Vaughn Gun.

the piston in the mud cylinder to eject the clay on every stroke. The gun is supported by a pivoted bracket, by which it may be swung into place and clamped before discharging.

When the shutting of the tapping-hole is completed, the blast is put on full and smelting proceeds. Iron and cinder accumulate in the hearth. After about two hours, the cinder arises again to the tuyeres and must be flushed. This is accomplished by simply drawing the plug from the cinder notch and allowing the cinder to run until it is drained to the level of the notch. The hole should be kept free during flushing by means of a pricker. After the first flush following a cast, it is necessary to flush again as soon as the cinder rises to the tuyeres. The period is much shorter than before, owing to the accumulated iron in the hearth. A third flush and perhaps a fourth takes place before tapping A sample of cinder is taken every flush, in a way similar to the sample of iron. It is needed for inspection and analysis. Usually the average of several flushes is analyzed.

CHARGING THE FURNACE.

When the furnace has been filled and starts on its career of reduction and melting, it tends to empty itself rapidly as the stock sinks. Constant vigilance must be exerted to keep a full column of materials. Proper filling and distributing of stock are of vital importance.

Since the furnace charges must be determined beforehand, and used in the proportions thus predetermined, it is necessary that the different kinds of stock should be carefully weighed. The stock house must therefore be provided with suitable scales. The use of a multiple beam scale is universal. This enables a separate beam, properly counterweighted, to be set aside for each kind of stock, and the weigher has only to see that each barrow balances on the proper beam before sending it up.

The basis of the furnace charge is the weight of fuel in each "round." Usually a definite number of pounds of fuel, varying from 4,800 to 14,000, according to the capacity of the furnace, is taken as the basis of the round and this quantity remains fixed. The proportions of ore and stone, however, are variable. The ore is varied according to the heat development in the hearth of the furnace, and the stone is varied according to the requirements of the ore and fuel. The usual order of dumping is to put the charge of fuel in the first hopperful, and the ore and then the stone in the second. In the case of furnaces which are filled automatically ·by means of a self-dumping skip, the skip is so large that four skipfuls constitute a round. In the case of hand-filled furnaces, however, the weight on a stock buggy must not exceed the power of the man who handles it. Hand buggies, therefore, usually carry 1,000 to 1,800 pounds of stock. Coke is so light that it is customary to have larger buggies in order that too many may not be needed. A coke buggy rarely carries over 960 pounds and the number needed to constitute a round will vary from 6 to 12, according to the size of round adopted. It is usual to distribute the ore charge through an equal number of the smaller ore buggies, and half that number of buggies will suffice to hold the limestone. Usually the number of buggies to the round is kept constant, any variation in the furnace charge being made by varying the weight of ore and stone in each buggy. Since the multiple-beam scale permits the devotion of a separate beam to each kind of stock, and the counterweight may be clamped to the beam and the beam box locked, the weigher needs no discretion beyond selecting the right beam and seeing that the proper number of buggies to the charge goes up promptly.

In the case of automatically charged furnaces which are provided with stock distributors, the weighing, hoisting and dumping of the charges are in the hands of one or two men who operate from the stock house. In the case of hand-filled furnaces, the stock is brought to the scales by the bottom-fillers, weighed, run on the hoist, and sent to the top of the furnace, where it becomes the care of the top-fillers. The top fillers run the buggies to the hopper and dump them around the bell. After each round or definite fraction thereof has been dumped, the bell is lowered and the stock slides into the furnace.

Stock Distribution.—The proper distribution of the stock around the hopper is essential to a well-working furnace. In the case of automatic distributors, the distribution depends upon the type of distributor used. In the case of hand-dumping, the distribution depends upon the top fillers. It should not be left to their discretion, but a definite system of dumping should be adopted and rigidly followed. The basis of proper distribution is symmetry. The hopper should be divided into a definite number of dumping stations, as four, six, or eight; the charges sent up in a definite and constant order, and dumped symmetrically. If the first four buggies of coke are dumped at 1, 3, 5, and 7, the next four should be at 2, 4, 6, and 8. If stone be dumped first at 2 and 6, let the next be at 4 and 8. If one buggy of mill cinder goes in each round, let it be dumped in a spiral form, 1, 2, 3, 4, 5, 6, 7, 8, thus giving symmetrical distribution.

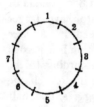

As we shall see later, the less the height of the column of materials within certain limits, the greater the need of fuel to maintain full furnace temperature. It is imperative therefore that the furnace should be kept full to the stockline, leaving only room enough for the proper manipulation of the bell. It is the duty of the top fillers to watch the height of the stock by gauging it at intervals. This consists of measuring the distance to the stock by means of an iron rod, thrust in turn through four small equidistant holes left for the purpose around the hopper in the furnace top. The stock should be gauged at all four points in order to discover whether it descends evenly over the whole area of the furnace. Irregularity of descent may not be detected by observing the same spot constantly.

OPERATION OF STOVES.

The care of hot blast stoves should never be trusted to unskilled hands, as thereby much damage may be done. In changing stoves, which is usually done every hour, the new one must be put on blast before the old one is taken off, in order that the blast may be continuous. In order to put the blast into a stove which is burning gas, it is necessary first to shut off the gas and to close the gas and air ports, and also the valve to the chimney, thus making the stove perfectly tight. Then the cold blast valve should be opened very gently until the stove is filled with the blast. This is done in order that the temporarily increased outlet for the blast may not cause the blowing engine to race. The hot blast valve may then be opened and we have two stoves on blast   The old stove may then be taken off in practically the reverse order   The hot blast valve is closed first, then the cold blast valve. The trapped blast must then be let out by cautiously opening an air port. When the pressure is relieved, the chimney valve and then the gas valve is opened, and the air ports regulated to suit conditions.

When the blast is off the furnace, no gas comes through the downtake to the stoves. Meanwhile the chimney is drawing cold air through the stoves on gas and robbing them of their heat.   Hence the gas and air inlets should be closed whenever the blast stops. If the stop is to be of considerable duration, the chimney valves should be closed and everything made tight. The hot and cold blast valves of the stove in use should be closed also.

If a stove gets cold through over use or from any other cause, and is not hot enough to ignite the gas upon its entrance, burning wood or waste should be placed in the combustion chamber to insure ignition and prevent explosion. The quantity of gas that may be burned in a stove is ordinarily all that can be spared from the boilers   It will vary with the kind of gas. The air ports must be regulated to satisfy the requirements of the gas in order to bring about complete combustion. Any excess of gas passes through the stove unconsumed, and any excess of air passes through unchanged. In either case no heat is given up to the stove but some is taken away, since any useless excess of either absorbs heat and carries it out of the stove

Cleaning the Stove.—Dust, which is brought to the stoves by the gases, wall accretions, pieces of brick, etc, fall and accumulate in the bottom of the combustion chamber in a more or less fused condition. It is necessary that the cleaning doors should be opened and this dirt removed every few days   Some of the dust is also carried up through the combustion chamber, falls through the checkers, accumulates in the chambers beneath, and gradually chokes up the passages. At least once a month all of the cleaning doors should be opened to scrape this dirt out. Semi-fused incrustatons adhere to the top of the checker flue walls, where they gradually bridge across and stop the flues. Such accumulations should be removed at least every six months, by letting the stove cool down and opening the cleaning doors on the top.

INTERRUPTIONS IN WORKING.

When once blown in, a furnace usually continues in blast for some years, until some portion, generally the lining, is damaged beyond repair, when it is " blown out." There are, however, two forms of temporary interruption, known respectively as " banking " and " blowing down "

Banking the Furnace.—It is frequently desirable to suspend temporarily the operation of a furnace, owing to lack of supplies, labor troubles, extensive repairs, etc. In such cases the furnace is " banked " by stopping the blast and smothering the fire by filling with clay all openings which would admit air to support the combustion of the fuel. During the period when the furnace is banked, it will lose heat by radiation, in cooling water, etc, which will leave a deficiency of heat in the hearth when the blast is put on again. To provide against this deficiency it is necessary to precede the stop by a large blank of coke, followed by light charges of ore not exceeding one-half to two-thirds normal, depending upon the length of the stop expected. The longer the stop is to be, the larger should be the coke blank and the lighter the subsequent charges of ore. At the same time, a blanket of fine ore spread over the top of the stock tends to seal the body and prevent draught. As soon as the coke blank has reached the bosh, the blast should be taken off and the furnace tapped clear of iron and cinder so that there will be nothing left to solidify. The tuyeres and cinder notch are then removed, the openings bricked up and every crevice from the mantle down, luted with clay and the gas burners of the stoves and boilers closed to prevent draught. In spite of these precautions, some combustion will take place and the stock will sink in the furnace. On starting up again, it is advisable to fill with coke the space due to settling, and then to start light, but gradually increasing, charges of ore.

The history of a successful banking charge of an 80 × 15 foot furnace, making basic iron, was as follows:

Fuel blank	. . . .	165,000 pounds coke.
Flux ..	. . . .	32,500 pounds limestone.
Charge	{	27,500 pounds coke. 28,000 pounds ore. 12,000 pounds stone.
Blanket		24,000 pounds fine ore.
Charge	{	27,500 pounds coke. 28,000 pounds ore. 12,000 pounds stone.
Fuel blank		27,500 pounds coke
Blanket		24,000 pounds fine ore.

The furnace was banked ten weeks, during which the stock settled 20 feet, an average of 2 feet per week. This space needed 66,000 pounds coke to fill it and that weight represents approximately the amount of coke burned in ten weeks, through infiltration of air. When the furnace was opened up, at the end of ten weeks, the crucible was full of ashes and good coke was before the tuyeres. The ashes were raked out through both notches and the notches closed, the tuyeres put in and the blast put on. The first cinder, flushed 6 hours later, was hot and gray, having 31.2 per cent. SiO_2 and 19.8 per cent. Al_2O_3. The first iron was tapped at the end of 24 hours, and was high in silicon and sulphur, having Si., 2.26 per cent. and S., 0.386 per cent. On the fourth cast the analysis was Si., 0.88 per cent and the S., 0.049 per cent. By the sixth day the furnace was in normal condition, making iron that averaged Si., 0.80 per cent and S., 0.028 per cent., and cinder that averaged SiO_2, 33.1 per cent. and Al_2O_3, 13.8 per cent.

If the suspension is expected to last only a few days, the coke blank may be reduced to a half or three quarters of the above amounts, but the other details may be practically identical. When the exigency is so sudden that time is not given to prepare the furnace for a stop, the only course is to remove the tuyeres and cinder notch and plug all the openings at once. Before starting up, however, it is advisable to open the notches and take time to cut through all chilled material to clean stock and to fill the openings with a mixture of sand and fire clay. This precaution will ensure the ready opening of the notches when needed. If it is found that a solid crust has formed across the hearth beneath the tuyeres, the explosion of a few sticks of dynamite in the centre of the hearth will break up the crust so that molten material may reach the bottom. If this precaution is neglected, it may be necessary to tap the iron through the cinder notch for several days.

When the blast is put on and charging begins, a coke blank, followed by light charges should be used first, in order that the hearth may have an opportunity to regain its lost heat as soon as possible. For a furnace 80 × 15 feet, a coke blank of 30.000 pounds should be used, followed by ten charges of ore, which are about one-third normal. This should be followed by another coke blank of 20,000 pounds, upon which gradually increasing charges may be built up, as before.

Blowing Down.—When a furnace is going out of blast and also in some instances of irregular working, charging is stopped and the stock is allowed to settle with gradually decreasing blast. This phase of operation is termed "blowing down." When blowing down is merely a corrective measure, it rarely proceeds as far as the bosh before its purpose has been accomplished. As the charges sink without the addition of cold stock, the top of the furnace gets hot and explosions are very liable to occur. To keep the top cool and to prevent warping the bell or hopper, some method of cooling is necessary. The most satisfactory way is to use sprays of water, and to keep the temperature between 500 and 600 degrees F. This is done by using perforated inch pipes inserted in the furnace top through the four gauge holes and projecting to a depth of 5 feet below the bell. The finer the sprays, the greater is their efficiency. The refilling of the furnace should be preceded by a coke blank, followed by light charges as before.

Blowing Out.—When conditions are such that it is desirable to end the campaign of a furnace, the furnace is "blown out." The first phase of blowing out a furnace is identical with blowing down. As the stock sinks, sprays of water are used to keep the top cool. As the body of material gets thinner and lighter, the pressure of the blast is lessened. The blast may continue until the top of the stock is within a few feet of the tuyeres, when it is necessary to stop because even a light blast will blow the stock about in the empty furnace without any further advantage. When the blast is finally stopped, all of the tuyeres and coolers are removed from the furnace, the remnants of the stock are raked out of the hearth and the furnace is ready to be dismantled for repairs.

CHAPTER IV.

BURDENING THE FURNACE.

The Furnace Burden.—The ratio of the ore, with its accompanying flux, to the fuel of a furnace charge is generally termed the "burden" of the furnace. The task of determining the quantity of each which is best suited to furnace conditions is designated as "burdening the furnace." The successful running of a furnace probably depends more upon proper burdening than upon any other single factor in its management. In the early days of the industry, before the constant application of chemical analysis to the materials used, burdening was a combination of previous experience and guesswork. If any raw material from an unfamiliar source had to be used, the treatment required had to be guessed until it could be determined by experience. With the application of analytical methods, however, it became possible to predict with tolerable accuracy the requirements of any materials from their chemical compositions.

Furnace Control.—The constant care of the furnaceman must be centered chiefly upon two aims; to keep the furnace working freely, and to maintain a satisfactory product. The two variables at his command, by which he must achieve results, are heat and slag composition. With sufficient heat and properly composed slags a wide range of products may be obtained from identical materials. The proper amount of heat in the furnace is that which will maintain the least temperature necessary to perform the work desired. Any excess is waste. The temperature which is necessary to attain given results depends upon the kind of product desired, and is inseparably linked with the slag composition. Generally speaking, the slag is the substance that requires the highest temperature for its proper disposal, and when that temperature has been attained it is ample for all other considerations. The amount of heat needed for the proper behavior of slag is dependent chiefly upon the slag composition.

SLAG.

By the uninitiated, slag is too often looked upon as a highly undesirable, but quite unavoidable excoriation from metallurgical processes. The proper mental attitude toward a slag is to consider it a reagent divinely appointed for the purification of metals. Since in any metallurgical process, all of the non-volatile constituents must appear in either the metal or the slag, it follows that whatever we would eliminate from the metal must be accommodated in the slag. This result can be accomplished only by giving to the slag such a character that it will offer to the impurity a stronger attraction than is offered by the metal.

Slag Constitution.—The cardinal difference between a metal and a slag determines whether a given particle shall enter the one or the other. This difference lies in the fact that the slag is composed essentially of elements which are oxidized, while the metal admits them only when in the elemental condition. Silicon, manganese, phosphorus, sulphur, and iron may exist simultaneously in both slag and metal. In the slag, silicon will exist only as SiO_2, but in the metal as Si. In the same way, all of the MnO, P_2O_5, SO_2 and FeO will be found in the slag, while the metal will be found to contain elemental Fe, Mn, P and S. Since all the elements are in the condition of oxides when they enter the furnace, it follows that they can be found in the metal only after they have lost their oxygen. Any elements, however, which are not deoxidized by the action of the blast furnace can therefore never be found in pig iron. This is why Ca, Mg, Al and alkalis are always absent from pig iron although present in the furnace in abundance.

Blast furnace slags consist primarily of calcic silicate, although either the lime or the silica may be partially replaced by other radicals. The lime is often partly replaced by magnesia, and generally by small quantities of the oxides of iron, manganese and the alkalis. The silica is always accompanied by alumina, and also by sulphur, either in the oxidized or elemental condition. The vast majority of blast furnace slags to-day will fall between the following limits in composition:

	Per cent.			Per cent.
SiO_2	25 to 50	MnO		Tr. to 2
Al_2O_3	5 to 20	K_2O }		Tr. to 1
CaO	25 to 50	Na_2O }		Tr. to 1
MgO	Tr to 25	S		Tr. to 2
FeO	Tr. to 1	P		Tr.

While it is probable that there are at least traces of all of the above constituents in every blast furnace slag, yet all but the first three might be absent without interfering in the least with its functions.

Slag Function.—The function of the slag is the removal from the furnace of any non-volatile matter that does not properly belong in the pig iron. We have seen that the alumina and all of the earthy and alkaline bases naturally enter the slag, and that they are accompanied by the bulk of the silica and sulphur. If we neglect for the moment the unavoidable traces of iron or manganese, it becomes evident that the only variables in this list are the silica and sulphur, and it is in respect to these two that the functions of the slag are chiefly manifested. The removal of the sulphur and the regulation of the silicon in pig iron are tasks which devolve largely upon the slag.

A suitable blast furnace slag is always characterized by perfect fluidity at furnace temperatures. All constituents must be in complete fusion so that all the fluxing power will be exerted to the utmost and in order that the slag may be readily drained from the furnace. Such a slag will have a quiet, rapid flow without froth or viscosity. "Hot, fluid and gray" is a time-worn description of good slag which still holds true.

Slag Efficiency.—The efficiency of slag in controlling the quantity of silicon and sulphur in pig iron depends upon its ability to take up and retain these two elements. This ability in turn depends upon the basicity of the slag. Properly speaking, the term "basicity" refers to the proportion of bases in a compound, but in the case of the blast furnace slag the basicity is usually reckoned in terms of the acid constituents—viz., silica and alumina. That is to say, it is the custom, as a rule, to watch the percentage of silica and alumina in the slag from day to day, and to disregard the percentage of bases. The result is the same in the end, however, for whatever is not acid in the slag must be base.

Slag Composition.—While the SiO_2 and Al_2O_3 together are usually considered as the acid constituents of the slag, they are not by any means co-ordinate. SiO_2 is always a strong, definite acid, no matter where it is found. Al_2O_3, on the other hand, is more often found acting as a base. It unites with all acids to form aluminic salts much as other bases do. It even unites with SiO_2 itself to form clay, an aluminic silicate, in which it performs unaided all the duties of a base. But in the presence of other more positive bases and particularly when there is a deficiency of acid constituents, the dual nature of Al_2O_3 is manifested, and it cooperates to supply the deficiency. It is under these conditions that it exists in the blast furnace slag, which may be considered usually a tolerably basic compound. The power of Al_2O_3 to act as a substitute for SiO_2 in slags is forcibly illustrated by the ease with which highly aluminous slags give up silicon for reduction and incorporation with the pig.

Acid Constituents.—The united percentages of SiO_2 and Al_2O_3 in the blast furnace slag usually range between 40 and 50 per cent. of the whole, though occasionally higher. They cooperate with such harmony that the Al_2O_3 ranges quite generally from 5 to 20 per cent. without materially altering the nature of the compound. It has been shown by Howe that in all silicates, ranging from subsilicates to trisilicates, in which the lime ranges from 30 to 45 per cent, substituting Al_2O_3 for SiO_2 does not materially affect the heat of formation. It should not be understood, however, that the resemblance is exact. As would be naturally expected from its neutral nature, the substitution of Al_2O_3 for SiO_2 without altering the bases, will give a slag of more limey appearance and greater basic power for the same total acids, than a higher proportion of SiO_2 to Al_2O_3 would give, yet without undue refractoriness. When Al_2O_3 is in excess of SiO_2, their combined amount may rise to 60 per cent. without losing the characteristics of a basic slag. On the other hand a highly aluminous slag, which is deficient in bases will not reveal its acid properties so markedly as when the proportion of SiO_2 is higher. It is said, however, that the substitution of 20 per

Tr. A. I. 1898

cent. MgO for an equivalent quantity of lime will produce viscosity in slags having over 10 per cent. Al_2O_3.

As a rule $SiO_2 + Al_2O_3$ amounts to 45 to 48 per cent. of the slag, of which about one-third is Al_2O_3 and two-thirds SiO_2. The percentage of the slag which they represent is regulated by the proportion of stone used for flux. The ratio which they bear to each other, however, depends upon their original proportions in the materials of the charge, which cannot be easily regulated. In the majority of ores the proportion of Al_2O_3 to SiO_2 in the gangues is about 1 to 4 or 5. The ash of the coke, however, being generally of slaty composition, and therefore akin to clay, is generally high in Al_2O_3. This fact, coupled with the removal of SiO_2 through deoxidation, accounts for the final ratio of 1 to 2 in the slag. Some average ratios of SiO_2 to Al_2O_3 in slags made from ores of different localities in United States are given below:

Locality.	SiO_2	Al_2O_3.	$SiO_2 + Al_2O_3$.	Ratio of SiO_2 to Al_2O_3.
Lake ores, P. S. Co.	30	17	47	1.8
Lake ores, C. I. & S. Co.	32	14	46	2.3
Cornwall ore	34	15	49	2.3
Alabama ore	35	15	50	2.3
Virginia ore	38	9.5	47.5	4.0

In addition to the SiO_2 and Al_2O_3, sulphur acts as an acid radical and unites with lime to form CaS. It is probable that this union cannot occur except in the presence of carbon or other reducing agent, which can unite with the oxygen given up by the lime as in the following reaction:

$$CaO + C + S = CaS + CO.$$

Each per cent. of sulphur neutralizes 1.25 per cent of Ca in the slag, and the calcic sulphide so formed, exists dissolved in the silicate of the slag. The quantity of sulphur in the slag ranges usually from 1 to 2 per cent. The solubility of the sulphide in slags is low, hence but little is dissolved and it separates readily. The solubility increases with temperature and basicity. The presence of sulphur in iron and slag simultaneously may be looked upon as a two-phase solution, and the distribution of sulphur between the two phases depends upon the composition of each. The coefficient of solubility in slags is increased by basicity, and that in iron by the presence of carbon and manganese. Evidently, therefore, absolute desulphurization of iron by slag is impossible.

Basic Constituents.—The other half of the slag is made up of bases. The ratio of bases to acids in slags usually ranges from 1 to 1.3, although these limits are often exceeded. In a large percentage of successful slags, the sum of the earthy bases $(CaO + MgO)$ approximately equals the sum of the acids $(SiO_2 + Al_2O_3)$, each being about 47 per cent., while the remaining 5 or 6 per cent. is made up of CaS, alkalies and oxides of the heavy metals. The latter, however, under normal working, are fixed by the unalterable conditions of the problem at each furnace, and are not subject to material manipulation. The proportions of the earthy bases, on the other hand, are under strict control and may be subjected to considerable variation. While the ratio of CaO to MgO in the slag may be affected to some extent by the composition of the gangue and the ash, it is mainly dependent upon their ratio in the flux used. Some limestones are practically free from magnesia and, in consequence, the slags which result will carry very little MgO. On the other hand, magnesian limestones are much used for flux in some localities. A pure dolomite contains 30.43 per cent. CaO and 21.74 MgO, a ratio of 1.4 to 1, which is about the maximum ever found in slags. It must be remembered, however, that MgO has a fluxing power 1.4 times that of CaO, which brings them on a parity in efficiency when in such proportions. Generally the percentage of MgO in blast furnace slags does not fall far below 5 per cent. and rarely goes much above 20 per cent. Under ordinary conditions it may vary between these limits without affecting materially the quality of the slag. As a rule the addition of MgO to calcareous slags lowers the melting points, since poly-basic silicates are usually more fusible than silicates of a single base. With tolerably pure silicate slags, having less than 5 per cent. Al_2O_3, the proportion of MgO need be of little concern, since successful slags have been run with CaO as low as 12 per cent. If Al_2O_3 exceeds 10 per cent, however, MgO in excess of 20 per cent. causes too much viscosity in the cinder, whereby it becomes sticky and does not work freely.

Fusibility of Slags.—Slags may best be regarded as mutual solutions of various oxides, and it is necessary for the more fusible components to fuse in order to dissolve the less fusible. For this reason the temperature of the formation of slag is generally higher than its melting point. The finer the state of division, and the more intimate the mixture, the nearer these temperatures approach. Experiments by Boudouard upon the melting points of various mixtures illustrate the mutual effect of different components upon each other. He places the melting point of SiO_2 at 3325 degrees F., and states that the addition of Al_2O_3 lowers the melting point. The minimum is reached with 15 per cent. Al_2O_3, which melts at 3075 degrees F. Further addition of Al_2O_3 causes a rise of melting point. 46 per cent. Al_2O_3 and 54 per cent. SiO_2 melts at the same temperature as pure SiO_2. $Al_2O_3SiO_2$ melts at 3435 degrees F and pure Al_2O_3 probably melts far above 3600 degrees F.

Inst. Jour., 1905, I., p. 339.

The addition of CaO to SiO_2 lowers the melting point rapidly. 30 per cent. CaO brings the melting point below 2730 degrees F , 40 per cent. gives the minimum melting point, 2650 degrees F. Further addition of CaO causes the melting point to rise gradually but irregularly, till at 90 per cent., it is again 2730 degrees F.

The addition of 13 per cent. Al_2O_3 to CaO gives a melting point of 2640 degrees F., 55 per cent, Al_2O_3 gives the lowest melting point, 2540 degrees F. Above 60 per cent. Al_2O_3 the melting point rises rapidly to that of pure Al_2O_3.

The following analyses by the same author show how little the melting point is affected by considerable changes in CaO and SiO_2:

SiO_2.	CaO.	Melts at degrees F.
62 0	38 0	2,590
51.8	48.2	2,625
34 8	65.2	2,660
21.2	78.8	2,660

The addition of Al_2O_3 to such a lime silicate, however, lowers the meeting point materially, thus:

SiO_2.	Al_2O_3.	CaO.	Melting point, degrees F.
34.8	0.0	65.2	2,660
37.4	10.3	52 3	2,515
40.0	22.7	37 3	2,450

Analyses by Gredt show clearly that the substitution of MgO for CaO in alumino-calcic silicates lowers the melting point until MgO nearly equals CaO, then raises it sharply as CaO is completely replaced by MgO, thus:

SiO$_2$.	Al$_2$O$_3$	CaO	MgO	Melting point, degrees F
40.30	23.05	33.87	2.69	..2,510
41 74	23 82	23 33	11 14	. 2,465
42 20	24 09	19 66	14 05	..2,465
43 68	24 03	8 14	23 25	..2,520
44.19	25 22	4.12	26.47	.2,570
44.72	25 52	0.0	29 76	...2,725

Inst. Jour., 1889, II, p. 413

That the melting points of slags are influenced primarily by their degree of basicity is shown by the following analysis selected at random from Boudouard's experiments:

SiO$_2$.	Al$_2$O$_3$	Total acids.	Total bases.	Melting point, degrees F.
54 8	0	34 8	65 2	.2,660
26.6	11 4	38 0	62 0	.2,625
23 9	20 3	44 2	55 8.	2,695
35 2	10 0	45 2	54 8	..2,625
37.4	10 3	47.7	52 3	.2,515
38 6	16.4	55.0	45 0	.2,500
40.0	22 7	62 7	37 3.	.2,450

Usually slags having more than 50 per cent. lime will fall to powder on cooling, owing probably to a change of volume which they undergo as the result of a molecular rearrangement.

Relation of Slag and Temperature.—The degree of basicity of the slag may be varied by varying the proportion of flux used. By this means two properties of the slag are changed, both of which tend to increase its effectiveness; its attraction for silica and sulphur is increased and its melting point is raised. The most fusible slags have the acid constituents somewhat in excess of the basic, and therefore any increase of bases tends to decrease fusibility. On the other hand, a large excess of silica tends to make an infusible slag, but the infusibility increases more rapidly by addition of lime than by increase of silica. The control of the hearth temperature and consequently the grade of iron is largely dependent upon the melting point of the slag. It is not easy to get the hearth temperature very far above that point, since the excess of heat is absorbed in super-heating the slag without affecting the iron. A slag with a low melting point will not be accompanied by hot iron, no matter how great an excess of fuel is used. This is best accounted for by the fact that the layer of slag prevents the absorption of heat by the iron. As a result the temperature of the iron is largely acquired from contact with the slag as it drips through it. A slag which melts readily is not long in the zone of combustion and does not acquire a high temperature. A refractory slag, on the other hand, passes into fusion slowly and becomes super-heated by long contact with the heated gases and therefore has the power to super-heat the iron on its passage to the hearth. This serves to explain the well-known but badly expressed fact that "lime gives heat to a furnace."

Relation of Slag and Product.—In general it may be stated that, other things being equal, a high hearth temperature will produce a high silicon and low sulphur product, and a basic slag will make both silicon and sulphur low. The usual explanation is that high temperatures facilitate reduction of silicon but tend to volatilize sulphur, while the basic slag holds them both in check. The two conditions act concurrently in restraining the sulphur, but are opposed as regards the silicon. It is by taking advantage of these tendencies that the control of silicon and sulphur is effected. Let us assume by way of illustration, a few examples:

1. High temperature and basic slag; $SiO_2 + Al_2O_3$ below 45 per cent. This would result in making both Si and S low in the pig, probably less than 1.00 per cent. and 0.04 per cent., respectively, hence it would be suitable for basic steel melting.

2. High temperature and neutral slag; $SiO_2 + Al_2O_3$ about 47 to 48 per cent. This condition would result in increasing both the Si and S in the pig, the former much more than the latter, since the high temperature still favors reduction of Si, and it is less strongly held in the slag. The S is affected by the change in slag also, and the resulting pig might run 1.5 per cent. Si and 0.05 per cent. S, which would be of Bessemer and forge grades.

3. High temperature and acid slag; $SiO_2 + Al_2O_3$, exceeding 50 per cent. This would probably result in a still further increase of silicon and sulphur for the same reason as given under (2).

The pig would probably analyse 2 to 3 per cent. Si, S under 0.06 per cent., and would be of foundry quality.

A decrease of temperature in any of the three cases under supposition would probably result in a lowering of Si and a rise in S. It is not possible to predict fixed results, from a given slag composition, however, as results are affected by so many conditions as to be always uncertain. These figures are given simply to fix the ideas by a concrete example of what might be reasonably expected. It has been shown by Howe that foundry and forge grades of iron usually accompany slags that are more basic, and white and charcoal iron accompanies slags that are Tr. A. I.
1896. less basic than singulo-silicates, in which the ratio of oxygen in the bases to that in the acids is 1.

Limit of Efficiency of Slags.—The limit of the power of slags of ordinary degrees of basicity to hold sulphur is probably not much above 2 per cent. of sulphur. As we have seen, the sulphur present in the charge per 100 pounds of pig is usually about 1 pound. In order that the sulphur in the slag may not exceed 2 per cent., therefore, it is necessary that there should be about 50 pounds of slag for every 100 pounds of pig made. In practice it is found that if the quantity of slag per ton of pig falls much below 1000 pounds, the extraction of sulphur suffers under any conditions of temperature or slag composition. It is evident, therefore, that with fuel of normal ash and flux of usual efficiency, the ore mixture should not contain less than an average of 8 per cent. of slag forming materials. If there is a deficiency of such constituents, it can be supplied by adding silica in some form and neutralizing it with flux. Silica may be added in several ways. The preferable source is a lean, siliceous ore, mixed judiciously with the richer ores. A very good substitute, however, is mill cinder, obtained from the heating furnaces of rolling mills. The cinder from a steel rolling mill is tolerably pure ferrous silicate low in phosphorus. That from a wrought iron rolling mill will be higher in phosphorus. The cinder from a puddle mill, however, should not be used unless a pretty high phosphorus pig is desired. Sometimes sand or siliceous rock is used.

It was formerly the custom to estimate the proportion of bases that should be in the slag for given results on the basis of the oxygen ratio. Identical oxygen ratios may give slags of very different ratios of acids to bases, when Al_2O_3 is reckoned as acid. It is found in practice that the efficiency of slag varies more nearly with the basicity than with changes in the oxygen ratio.

Physical Characteristics.—To the practised eye, the appearance of the slag tells much concerning its composition and the temperature of formation. A slag that is high in earthy bases has a light gray to bluish granular fracture when cold. When high in lime, slags " slake," or crumble to powder soon after cooling. This tendency is retarded by the presence of magnesia. As the proportion of bases decreases, the slag shows a vitreous tendency —at first on the outer edges only, but as the acid constituents increase, it may become glassy throughout. The presence of alumina opposes this tendency, and gives a more earthy appearance than silica alone. Siliceous slags can be drawn out into fine strings just before solidifying, while basic slags are very short. Hot acid slags, when run into the granulating pit, froth up into light fluffy heaps, while basic slags sink quietly to the bottom. Other things being equal, the slag from a cold furnace will be more vitreous than from a hot furnace. This is because it is more siliceous since less silicon has been reduced from the slag and gone into combination with the iron.

A siliceous slag is more likely to carry an appreciable quantity of oxide of iron than a basic slag This is because unsatisfied SiO_2 will seize upon any unreduced iron that reaches the fusion zone more readily than when it is saturated with bases. For this reason slags from a cold furnace generally have a dark color and high specific gravity, particularly when improperly prepared material is projected into the hearth. Calcareous slags sometimes give a dark color when a furnace is in trouble without losing their earthy appearance or containing an undue proportion of iron. The dark color appears to be simply a stain, probably caused by finely divided carbon.

According to Vogt, blast furnace slags, when allowed to cool slowly tend to crystallize into definite mineralogical forms, varying with their compositions. Bisilicate slags show enstatite,

Age, 1904

augite, and wollastonite, according to the proportion of Ca and Mg Singulo silicates show olivine and melilite and other tetragonal minerals, according to the ratio of CaO to the oxides of Fe, Mn and Mg Pyroxene appears in more siliceous slags. Feldspar and free oxides, such as quartz, corundum and metallic oxides are never found. Sulphur always appears as a monosulphide of Ca, Fe or Mn.

CONTROL OF HEARTH TEMPERATURE

For the production of different classes of iron the silicon content must be controlled within rather narrow limits. For example, less than 1 per cent. of silicon is usually demanded for basic open-hearth grade of pig, but such an iron would be entirely unsuited for foundry purposes. On the other hand, foundry irons usually contain upwards of 3 per cent of silicon, but such an analysis would be prohibitive for basic irons. The chief factor in the control of silicon is hearth temperature, which in turn is determined largely by the slag composition With a given slag composition, however, the hearth temperature is capable of considerable variation. There are several ways in which it may be diminished if desirable. The most immediate effect may be obtained by using a proportion of cold blast. Increasing the quantity of blast would give the same effect, and instead of checking the furnace would increase its output. This is equivalent in its effect to adding cold blast, since the increased quantity of air passing through the stoves does not permit each particle to be heated to so high a temperature. The effect is temporary, however, and the furnace soon comes to equilibrium on the new basis. This method is not usual now, however, as it is customary to blow a constant quantity of air per minute without variation. If the rate of driving cannot be increased, the cooling effect may be obtained by increasing the burden, which is equivalent to a diminished fuel consumption. This method is slower in its effect since no result is apparent till the new burden reaches the hearth. The reverse of these conditions, namely, decreased burden, or increased blast temperature, will result in increased hearth temperature.

The problem of burdening a furnace properly presents two distinct phases, which may best be designated by the titles of theoretical and empirical. The **theoretical phase** presents itself when one is dealing with new problems, such as unfamiliar materials, whether ores, fuel or fluxes, or when using familiar materials in unfamiliar proportions, such as in filling a furnace, preparatory to blowing in. The **empirical phase** is present when a furnace is operating with fixed sources of supply In other words, under strange conditions materials must be used according to analysis, and the proper proportion determined by calculation, but after approximately proper conditions with fixed materials are once attained, the diurnal variations may best be watched and corrected by simple inspection, checked by regular analyses.

THE THEORETICAL PHASE.

Limitations of Slag Control.—In order to obtain a slag of an approximately given composition from a given set of materials it is necessary first to know definitely the composition of the materials themselves, and then to arrange and adapt them to produce the required result. Since each member of a furnace burden contains gangue elements in tolerably fixed proportions, it is not always possible to produce results that are too closely limited. For instance, it would be manifestly impossible to produce a slag having an alumina-silica ratio of 1 to 10 when the ratio in the raw materials is 1 to 4. By grouping the Al_2O_3 and SiO_2 together, however, as the acid portion of the slag, and treating them as a unit, the difficulty of calculation is lessened without in any wise affecting the unalterable conditions of the problem. As a rule, the SiO_2 and Al_2O_3 of Lake Superior ores are so proportioned that an average ore mixture will approximate the ratio, SiO_2 to $Al_2O_3 = 4$ or 5. Resulting slags, however, generally show a ratio of about 2 to 3, due to the addition of Al from fuel and stone, and subtraction of Si by the pig iron.

Slag Calculation.—For the purpose of illustrating how such diverse materials may be arranged and combined to give definite results, let us assume the following composition of materials in per cent. or pounds per 100 pounds:

Material	SiO_2	Al_2O_3	CaO.	MgO	Mn	P.	S	Fe	C
Fuel	5.30	3.00	1.00	0.70	.	0.020	0.75	...	86.5
Ore	7.49	0.81	0.15	0.12	0.45	0.024	.	55.75	...
Flux	0.64	0.32	54.20	0.35	..	0.004	...	...	...

Let us assume that the resulting pig will contain 3.4 per cent. C, 1.5 per cent Si and 94 per cent Fe, and that the slag will contain 47 per cent. $SiO_2 + Al_2O_3$, and 51.5 per cent. bases, a ratio of acids to bases, 1 to 1.1

The problem is simplified by dividing it into three steps, as follows:

1. Find the available base and slag-forming constituents of the flux.

2. Find the flux needed and weight of slag formed by the fuel and its available carbon

3. Find the flux needed, weights of slag formed and fuel consumed by the ore.

With these figures it is easy to find the total flux required, and amount of slag formed, and also the probable phosphorus content of the resulting pig. If the flux requirement for each ore is expressed in pounds per hundred of ore, it greatly simplifies the task of changing the burden of the furnace whenever desired.

Flux Requirement.—In the case of the flux, the available base in the stone is as follows:

Acids.		Bases.	
SiO_2...	...0.64	CaO..........	54.20
Al_2O_3........	0.32	MgO ..	0.35
	0.96		54.55
	1.1 slag ratio.		1.05 neutralized by acids
	0.96		53.50 **available base.**
	0.96		
	1.056 bases needed		

$$\frac{100.0}{53.5} = 1.87 = \text{efficiency of flux.}$$

The non-volatile, irreducible parts of the flux which make up the **slag-forming constituents** of the stone are $SiO_2 + Al_2O_3 + CaO + MgO$, and they equal 55.5 per cent. of the stone.

Fuel Requirements.—In the case of the fuel, the flux needed, the slag formed and the available carbon may be found as follows:

Acids.		Bases	
SiO_2 ...... 5.30		CaO 1 00	
Al_4O_3 .. .3 00		MgO. 0 70	
8 30		1.70	

1 1 slag ratio.

    8.30
    8.30

9.13 bases needed.
1.70 bases present

7.43 bases to be added.

$7.43 \times 1\,87 = 13\,89$ pounds stone needed to flux ash

$0\,75 \times \dfrac{56}{32} \times 1.87 = 2.45$ pounds stone needed to flux sulphur.

Total   .   ...16 34 **pounds stone needed per 100 pounds fuel.**

The slag formed by the ash and sulphur of the fuel equals the ash and sulphur plus the flux needed, thus:

$8.3 + 1.7 + 0.75 + (16\,34 \times 0\,535) = 19.81$ pounds, **weight of slag per 100 pounds.**

$19\,81 \times 0\,25 = 5\,0$ pounds carbon needed to melt slag

$86\,5 - 5.0 = 81\,5$ per cent. **available carbon.**

Ore Requirements.—In the case of the ore, the flux needed and slag formed may be found as follows:

Acids.		Bases.	
SiO_2.........7 49		CaO... ....0 15	
Al_2O_3........ 0.81		MgO... ..... 0 12	
		¼ MnO 0 20	
8 30		0 47	

Less 1 80 reduced to Si

6 50 to be fluxed
1.1 slag ratio.

    6 50
    6 50

7 15 bases needed
0 47 bases present.

6.68 bases to be added

$6.68 \times 1\,87 = 12\,49$ **pounds stone for 100 pounds ore.**

$\dfrac{94.00}{55\,75} = 1.686$, number tons ore to make 1 ton pig.

$1.686 \times 0.1249 = 0\,2106$ **tons stone per ton pig.**

The slag formed by the ore per ton of pig may now be found by simply adding the slag forming constituents of the ore to those of the stone needed to flux it, thus:

$\dfrac{6.5 + 0.47}{100} \times 1\,686 = 0.1175$ tons slag from the ore.

$$\frac{55\,5}{100} \times 0.2106 = 0\,1169 \text{ tons slag from the stone}$$

$$\overline{0\,2344} \text{ total slag due to ore}$$

Fuel Consumption.—Having found the quantity of slag formed by the ore and its flux per ton of pig, we may ascertain the fuel requirement per ton of pig by simply adding that needed by the slag to that needed by the pig, as follows:

For the formation and melting of the slag, $\frac{0\,2344}{4} =$. ....0 059 parts C.

For reduction, impregnation and melting pig containing 1 5% Si = 0 685 parts C.

Total carbon required for pig and slag ....0.744 parts C.

$$\frac{0\,744}{81\,5} \times 2240 = 2045 \text{ pounds coke per ton of pig}$$

Allowing 6 per cent for braize, the fuel consumption per ton of pig will be about 2150 pounds

Total Slag.—The total quantity of slag per ton of pig will be as follows:

Pounds
From the ore, 0 2344 × 2240 =......................525.0
From the fuel, 0.1981 × 2045 =......................405 1

Total slag per ton of pig =......................930.1

Total Stone.—The total quantity of stone per ton of pig is as follows:

For the ore, 0.2106 × 2240 =.......471.7 pounds
For the fuel, 0.1634 × 2045 =.......334 1 pounds.

Total stone per ton of pig =........805.8 pounds, which is 21.4 per cent. figured on the burden and 36 per cent. figured on the product.

Furnace Charge.—Based on a 5-gross ton charge of fuel, the rounds of the charge would be composed as follows:

Pounds
Fuel 11,200
Ore 21,500
Stone 4,500

Phosphorus Content of Pig.—Assuming that practically all of the phosphorus in the charge enters the pig, the percentage to be expected there may be found as follows:

Pounds per ton pig.
From the ore, 0 024 × 1 686 × 22 4 =......................0 9060
From the fuel, 0 020 × 20 45 =......................0.4090
From the stone, 0 004 × 7 99 =......................0 0320

1 347

$$\frac{1.347 \times 100}{2240} = 0.06 \text{ per cent P.}$$

Pig = 94% Fe, 1.5% Si, 3.4% C

Flux = 50% available base and 5% of non-volatile substance .... P = .020

Slag : Acids equal bases. 1 lb C melts 4 lbs. Slag.

a — Fe in Ore dried @ 212° F. (%)	b — Ore needed for 100 lbs. pig for Fe (lbs)	c — Fe₂O₃ needed for 100 lbs pig (lbs)	d — Gangue per 100 lbs pig (lbs)	e — Gangue to be fluxed (lbs)	f — Stone needed per 100 lbs pig for ore and fuel (lbs)	g — Wt. of Slag per 100 lbs pig from ore and fuel (lbs)	h — Weight Coke per 100 lbs pig (lbs)	i — Air needed to burn Coke per minute (cu ft)	j — Heating surface needed (sq ft)	k — HP needed 8 HP per ton Coke (HP)	m — Space occupied by ore per ton pig (cu ft)	n — Space occupied by stone per ton pig (cu ft)	o — Space occupied by fuel per ton pig (cu ft)	p — Space occupied by total stock per ton pig (cu ft)	q — Amount phosphorus in pig (%)	r — CO₂ of stone stolen by carbon per 100 lbs pig (lbs)	s — Carbon as CO₂ per 100 lbs pig (lbs)	t — Carbon as CO per 100 lbs pig (lbs)	u — Weight dry gases per 100 lbs pig (lbs)	v — Ratio CO/CO₂	w — Fuel consumption per ton (lbs)
66	142.4	184.3	8.1	2.47	4.95 / 24.40	7.62 / 24.60	87.1	29,800	157,000	2,800	17.00	6.00	69.70	94.2	.075	3.84	29.0	44.96	493.30	1.55	1,931
64	146.9	184.3	12.6	6.88	13.75 / 23.30	16.97 / 25.70	90.1	30,800	155,000	2,880	18.75	8.75	72.10	99.60	.080	4.44	29.0	48.50	508.05	1.67	2,018
62	151.6	184.3	17.3	10.64	21.90 / 29.44	23.80 / 33.8	92.9	31,830	160,000	2,970	19.60	10.60	74.30	104.5	.084	5.89	29.0	51.03	522.10	1.79	2,081
60	156.7	184.3	22.4	14.72	29.44 / 29.80	31.30 / 27.30	95.0	32,830	165,000	3,070	20.00	12.60	76.70	109.9	.088	6.40	29.0	53.51	537.00	1.92	2,148
58	162.1	184.3	27.8	19.04	38.10 / 29.60	45.5 / 28.8	99.2	33,920	170,000	3,170	21.65	14.75	79.30	115.7	.093	7.50	29.0	57.40	553.85	2.05	2,222
56	167.8	184.3	33.5	23.60	47.20 / 28.70	56.2 / 29.3	102.6	35,160	175,000	3,280	22.75	17.00	82.05	121.8	.098	8.66	29.0	63.39	570.90	2.19	2,298
54	174.1	184.3	39.8	28.04	57.30 / 29.80	68.1 / 30.4	106.4	36,420	184,000	3,400	24.00	19.50	85.10	128.6	.103	9.93	29.0	67.97	590.44	2.34	2,383
52	180.8	184.3	46.5	34.00	68.10 / 30.90	81.7 / 31.5	110.5	37,520	190,000	3,530	25.30	22.10	88.40	135.8	.109	11.27	29.0	73.20	600.51	2.51	2,475
50	188.0	184.3	53.7	39.76	81.70 / 32.10	94.2 / 32.7	114.8	38,900	196,500	3,670	26.80	25.00	91.80	143.6	.115	12.72	29.0	77.87	632.15	2.69	2,571
48	195.8	184.3	61.5	46.00	94.20 / 38.40	108.9 / 34.1	119.5	40,900	201,500	3,830	28.10	28.10	95.00	152.1	.122	14.30	29.0	83.48	655.90	2.88	2,677
46	204.4	184.3	70.1	52.88	105.75 / 34.10	125.1 / 33.5	124.7	42,700	213,500	3,990	30.10	31.50	99.70	161.8	.129	16.03	29.0	89.05	680.30	3.09	2,793
44	213.6	184.3	79.3	60.24	120.50 / 33.50	142.4 / 37.1	130.2	41,600	233,000	4,170	32.05	35.15	104.10	171.3	.137	17.90	29.0	96.67	709.05	3.30	2,916
42	223.8	184.3	89.5	68.40	138.80 / 37.10	161.5 / 39.9	136.4	46,700	243,500	4,360	34.15	39.20	109.15	182.5	.146	19.95	29.0	108.50	741.90	3.57	3,055
40	235.0	184.3	100.7	77.36	154.70 / 40.50	183.6 / 41.2	144.5	49,400	247,020	4,620	36.50	48.70	115.60	195.8	.155	22.23	29.0	112.60	772.70	3.88	3,237

(a) Assumed per cent of Fe in dried ore.

(b) $\frac{94}{a} \times 10$)

(c) $112 : 94 = 161) : X$

(d) difference between (b) and (c).

(e) [90% (d) − 3.2] − 16% (d).

(f) Twice (e) and 28% of (h)

(g) [(d) − 3.2] + 55% of 2 (e) and 28⅓% of (h).

(h) $\left[\dfrac{[(d) - 3.2] + 55\% (2e)}{4} + 66 \right] + .78$

(i) $\dfrac{22.4 (h) + 55}{96}$

(j) 5 (i).

(k) $\dfrac{22.4 (h)}{2240} \times 400 \times 8$

(m) $\dfrac{5 (c) + 2.5 (d)}{b} \times \dfrac{62.4}{1.7}$ divided into 22.4 (b). /

(n) 22.4 (f) + 100.

(o) $h \times \dfrac{22.4}{28}$

(p) sum of m + n + o.

(q) $\dfrac{64}{100} (b) + \dfrac{.02}{10} (f) + \dfrac{.015}{100} h$

(r) 11 % of (f).

(s) For a 58% ore may be found as follows:—

Oxygen in 184.3 lbs Fe_2O_3	40.80 lbs
" combined with 1.54 Si	1.71 "
" " Mn, P, S & C	0.99 "
Total oxygen to be extracted	43.00 "
Oxygen removed by CO (90%)	38.70 "
" " Solid Carbon	4.80 "

The carbon existing as CO_2 will therefore be the quantity which can unite with twice 38.7 lbs

O_2 or $\dfrac{38.7 \times 12}{16} = 29.0$ and the CO_2 formed equals 38.7 + 38.7 + 29.0 = 106.4.

1.5 Si = 3.2 SiO_2

(t) For a 58% ore may be found as follows :—

Carbon in the coke (85% of 99.2 lbs) =		84.32 lbs.
" stolen by stone (r) =	7.50	
" impregnated in pig iron =	3.40	10.90 "
" available		73.42 "
Required to remove 4.3 lbs. O_2		3.23 "
Left to be burned by the blast		70.19 "

70.19 lbs C requires 93.59 lbs N_2 to form CO which is accompanied by 308.83 lbs O_2 to form CO which is accompanied by 308.83 lbs N_2.
70.19 C + 93.59 O_2 (from air) + 98.70 O_2 (from ore) =
202.48 CO + CO_2, 202.48 − 106.4 = 96.08 CO
The total carbon existing as CO may now be found thus :

2 + 7.5 = 15.00	41.17 lbs C uniting to form	96.08 lbs. CO
8.23	" " "	85 Cu "
	" " "	7.53 " "
(f) 59.4		138.61

(u) For a 58% ore = 138.61 + 106.40 + 308.83 = 553.86. (See t & s)

(v) $= \dfrac{t}{s}$

(w) $= (h) \times 22.4.$

Other Methods of Calculation.—Methods have been devised for performing slag calculations mechanically by means of movable scales and by means of the slide rule, but their application is limited to the given conditions for which they are arranged, and they have not been adopted generally.

Inst. Jour., 1891, I , p 151.

Other methods of numerical calculation exist and have their advocates, but they show no advantage over the method used here. Graphical methods, when running for considerable periods, under fixed conditions, may have some advantages, however.

Tr A I M E, XXI., p 364

Iron Age, Dec. 3, 1891, Feb 3, 1892, Mch. 25, "

Effect of Ore Richness.—Influence of ore composition upon the consumption of fuel and flux and also upon the quantity of blast required, stove, engine and boiler capacities, gas volumes and carbon ratio is shown by the preceding table. It should be observed carefully that the relations are based upon thoroughly dehydrated ores. For a given iron content in the natural state the results will be lower in proportion to the amount of moisture present:

THE EMPIRICAL PHASE.

The foregoing statements constitute the gist of the theoretical phase of burdening a furnace. Such calculations are exceedingly useful in investigating the results which might reasonably be expected from a given mixture of raw materials, and the facts will probably come very close to expectations while everything runs smoothly. It is much safer, however, to judge the needs of a furnace by what comes out of it than by what goes in, and it is at this stage that observation and practical experience are of the utmost value. This phase of burdening may be denominated the empirical phase.

Watching the Furnace.—There are two cardinal points in the condition of a furnace which must be carefully watched, viz., the temperature of the hearth and the proportion of flux. If these two factors are right the furnace must work well. There are many causes for the interruption of proper temperature conditions, chief of which may be mentioned too heavy burden, leaking water blocks, improper combustion, improper blast quantity, irregular movement of stock. There are several indications by which the temperature of the furnace hearth may be judged with sufficient accuracy for practical purposes. It is not necessary to estimate the hearth temperature according to any absolute scale of degrees. It is necessary to judge only by comparison whether the temperature is above or below the point desired for the work in hand. The skill of the metallurgist is better manifested by knowledge of the temperature needed for certain results than by ability to detect fine differences in temperature.

. **Judging Hearth Temperature.**—The most direct means of judging the hearth temperature is by looking directly at it through the peep hole in the tuyerestock. If the hearth is cold it will show a reddish or lavender light, and the coke can be readily seen as it plays in the blast. If the hearth is very hot the light will be dazzlingly white, so that the coke may not be even discernible. Intermediate temperatures will present intermediate manifestations.

An indirect method of determining the temperature of the hearth is to observe the temperature of substances which come out of it. Chief of these are the iron and the slag. The temperature of the iron may be judged directly by its appearance or indirectly by its silicon and sulphur content. The iron has a high initial temperature when it runs from the furnace with a clear, rapid flow, free from scum or sparks, and does not readily skull or chill in the bottom of the runner. Such iron is likely to be fairly high in silicon. Cold iron usually sparks in the runner and has a scum, which may vary from thin, milky flakes, which dot the surface of the stream when slightly cold, to globules of solid iron, known as buckshot, when very cold and high in sulphur. Such iron usually chills in the runners before reaching the last pigs in the bed, and is apt to be low in silicon. On the other hand, some irons rich in graphite, though made in hot hearths, lack fluidity, and irons made in the presence of very basic slags may be very low in silicon and yet possess great fluidity, owing to a high initial temperature. It is evident therefore that in order to judge hearth temperature by the iron made in it, it is necessary to know something of the existing conditions and to take their influence into account.

The hearth temperature may be judged also by the slag produced in it. The absolute temperature of the slag that follows

the iron from a furnace usually appears to be much higher than
that of the iron which precedes it. Since the layer of slag is
nearer the tuyeres, it would naturally have a higher temperature
than the iron below it. Moreover, owing to the difference in
their respective specific heats, the slag contains nearly 50 per cent
more heat per pound than the iron does, and radiates it much
faster, thereby appearing to be hotter. The relative temperature
of slag may be estimated while it is flowing, from its color and
fluidity. Under ordinary conditions the whiter the light and
greater the fluidity, the higher the temperature. Hot slag usually
gives off a cloud of whitish fumes, which rises slowly from its
surface while it flows. The appearance of a piece of slag after
it has become cold may tell something about the temperature in
which it was made. Usually a hot slag will show a grayish white
color on the fractured surface after cooling, and will have a low
specific gravity. A slag made in a cold furnace, examined under
the same conditions, will usually be heavy and dark in color.

Another indirect means of judging the furnace temperature
is by observing the appearance of the flame of the waste gases
as they burn in the stoves or under the boilers. The gas which
comes from a cold furnace burns with a thin, bluish or lavender
flame, which has a clear, lambent appearance. The gas from a
hot furnace has a strong, yellow flame, which is opaque, and it is
usually accompanied by a whitish fume which increases in density
as the temperature rises.

Varying Hearth Temperature.—When the temperature indications show that a furnace is about to go " off her grade," that is,
getting a bit too cold, there are several ways in which it may be
righted. Sometimes a slight increase in blast temperature, such
as would result from putting on a fresh stove, will suffice. A
more positive effect may be obtained by somewhat reducing the
quantity of blast. A persistent coldness indicates insufficient fuel,
and consequently the burden should be reduced. If conditions
are such that a furnace becomes quite cold, it may be necessary
to charge a blank of fuel.

When a furnace works too hot, one of two courses may be
pursued to advantage. If the quantity of blast be increased, it
will melt more rapidly and make a larger output, thereby reduc-

ing labor costs per ton and fixed charges. If the blowing capacity
of the engines has been already attained, the other alternative is
to increase the burden, and thereby to reduce the costs of fuel
per ton

Effects of Cold Furnace.—If the furnace is allowed to get
cold with any kind of slag, the reduction of silicon may decrease to
the vanishing point. Whenever the silicon falls much below
1 per cent. of the pig with sulphur above 0 05 per cent., it is generally insufficient to cause the separation of much graphite, and
mottled or white iron results. As the silicon in iron decreases,
the sulphur generally increases. The change in sulphur is not
necessarily the result of the change in silicon content, but may be
partly attributed to it. There are three possible explanations of
the rise of sulphur, and it is probable that the total effect is the
result of all three. In the first place, it is claimed by some that
silicon has a distinct excluding effect upon sulphur and that in
consequence a high silicon iron cannot take up more than a few
tenths per cent sulphur. In the next place, the usual explanation of low sulphur in iron made in a hot hearth is that the intense
heat volatilizes a large part of the sulphur and conversely a cold
furnace will allow it to enter the pig.

Tr A I M E, XXIII., p 382

Thos Turner, "Metallurgy," p 200.

The final consideration, which is by no means a minor one, is
the decrease of basicity of the slag when the expected quantity
of silicon is not reduced from the siliceous components of the
charge. A very simple calculation shows that if a furnace is
making iron with 1000 pounds of slag per ton of pig, the transfer
of 1 per cent. of silicon from the iron to the slag in the form of
2.14 times as much SiO_2 will change the percentage of acids in
the slag from $47\frac{1}{2}$ to 50 per cent, and the ratio of bases to acids
from 1.1 to 1, which is equivalent to a decrease of over 10 per
cent. in the quantity of flux used A proportionate change in the
slag composition resulting from still less reduction of silicon,
would profoundly affect the desulphurizing power of the slag.

Under improper conditions of temperature and slag composition, insufficiently digested material is likely to descend into the
furnace hearth. In consequence a considerable quantity of unreduced oxide of iron may enter the slag to combine with the unreduced silica. A slag which is highly charged with ferrous sili-

cate has a low melting point and great fluidity, and exerts a strongly corrosive effect upon the lining or hearth accretions. It is consequently known as a " scouring " slag. With more regular and better watched conditions such a slag occurs less commonly now than formerly.

Another phenomenon which is peculiar to cold hearths is the occurrence of " buckshot." It appears in the form of unaggregated globules of iron which are intermingled with the slag that floats on the top of the flowing iron, or which are entangled mechanically in a cold and viscous slag. The globules appear to be chiefly iron which has been but partly carburized. They have undergone fusion and yet have been unable to coalesce successfully, owing to the resistance of the pasty slag.

Control of Manganese.—Like the silicon and sulphur, the manganese which enters the pig varies with the temperature and slag composition. Since a high temperature favors reduction of manganese, it follows that the hotter the furnace, the larger will be the proportion of the manganese present which will enter the pig. On the other hand, since oxide of manganese acts as a base, it is interchangeable with lime in the constitution of the slag. A highly calcareous slag, therefore, will tend to release more manganese for reduction than a siliceous slag. Under ordinary conditions it is safe to count on the reduction to metal of 50 to 75 per cent. of the total manganese present in the charge, and the remainder will enter the slag. Its power of uniting with sulphur and excluding it from the iron makes it desirable to have 0.5 to 0.75 per cent. Mn in the pig, which indicates that there should be in the furnace about 1 pound Mn for every 100 pounds of pig made, or about one-third per cent. of the charge.

Control of Phosphorus.—As stated before, the phosphorus content of pig iron is practically independent of furnace manipulation but depends almost entirely upon the nature of the materials used. The source of the phosphorus is in the fuel and flux as well as the ore, and that fact must be reckoned with in making the furnace charges. Probably never less than 90 per cent. of the phosphorus present enters the iron and more often it is nearer 100 per cent. It is safer, therefore, to assume that it will all enter the iron, particularly when a rigid phosphorus limit is required.

Control of Carbon.—The quantity of carbon which enters pig
iron is independent of the composition of the furnace mixture
and is not directly affected to any great extent by furnace conditions. The usual range of carbons in pig iron is from 3 to 4.25
per cent., but the great majority run from 3.25 to 4.00 per cent.,
and this quantity is tolerably constant, whether the iron is gray
or white. The color of the iron is determined by the proportion
of the carbon that exists as graphite and not by the total quantity
of carbon present. The total quantity of carbon present in pig
iron is profoundly affected by the presence of other elements in
the pig. Both silicon and phosphorus exert a strong excluding
tendency upon carbon, so that practically no carbon will be re-
tained by pig iron which contains 15 to 20 per cent. of either of
them. On the other hand, manganese has a strong attraction for
carbon and manganiferous irons are proportionately higher in
this element. The interrelations may be expressed approximately
by this formula, wherein the quantity of each element is expressed
in per cent.:

Total carbon $= 4.5 - 0.25$ Si $- 0.3$ P $+ 0.03$ Mn.

The variations of the different elements in gray and white
irons may be represented graphically as follows:

Titanium.—The presence of titanium in ores of iron had the
reputation of causing infusible and troublesome slags, due to the
presence of titanate of lime and other titaniferous compounds
That this reputation of titanium is unjustifiable is the claim of
A. J. Rossi. He states that titanates are not infusible, but that
in slags having about 60 per cent of acid constituents, as much
as 35 per cent. TiO_2 was substituted for SiO_2 without decreasing
the fusibility of the slag. More recent experience has justified
his claim

TiO_2 is not readily reduced by the influences in the blast
furnace, and therefore titanium is rarely found in pig iron.

Desirable Ores.—There are several points beside its composition which deserve consideration in deciding the desirability of
an ore In the first place, it should have enough slag-making
material so that the sulphur may not exceed 2 per cent. in the
slag The quantity of gangue for average ores should not be

E M J,
Sept 1, 1904,
p 350

Tr A. I. M. I.
XXI, p 832.

Iron Age,
Feb. 20, 1896,
p. 464

Min. Ind,
1900, p 715.

less than 8 per cent Secondly, it is desirable that the ores should lose their oxygen as high in the furnace as possible, so that little CO_2 will be generated below the point where it can attack carbon. Thirdly, a moderate carbon deposition is desirable to disintegrate the ore and facilitate reduction. An excessive deposit is to be avoided as it causes packing which results in excessive engine pressure and top explosions.

Condition of Pig.—Pig irons which are to be used for the manufacture of other metallurgical products, such as wrought iron or steel, by one of the processes of conversion, are always classed strictly on the basis of composition. If analysis shows that the various elements fall within the required limits, the physical appearance is usually considered quite secondary. With irons that are to be used without change of nature, however, as in castings, the physical appearance of the fractured surface is considered an important indication of the degree of suitability. The physical appearance follows closely the composition Other things being equal, the higher the silicon—up to 3 per cent. at least—the darker and coarser grained the iron. As silicon decreases, the fracture is lighter and closer until white is reached. Other conditions, however, may modify the effect of silicon. An iron made with a hot, limey cinder will present a grain whose openness is out of proportion to its silicon content, while high sulphur tends to make the grain close. The rate of cooling also modifies the appearance; the slower the rate, the larger the grain. It is for this reason that sand pig beds have been generally preferred to chills for foundry irons. The advantage is somewhat imaginary, however, as iron of the same composition which has been through the casting machine will make, when remelted, quite as good castings as the sand pigs. The following analysis of experiments by the Bethlehem Steel Company illustrates the effect of chills on iron:

	Sand pig.	Chill pig.	Sand pig remelted	Chill pig remelted
Si	3.000	2.990	2.910	2 950
Mn	0.950	0.950	0 850	0 840
P	0 770	0.773	0 769	0 764
S	0 041	0 011	0 061	0 071
G C	3 210	2 460	3 022	3.100
C C.	0 250	0.920	0 368	0.237
T. C	3 460	3 380	3 390	3.357
T. S.	15,000 pounds	41,000 pounds	16,300 pounds.	17,000 pounds

From the first two columns it is apparent that strength is not dependent upon silicon alone, but follows closely the condition of the carbon, no matter how that condition is produced. By comparing the results with the originals, it is evident that there is little or no change during remelting. Less "kish" is formed on chill pigs than on sand pigs. The higher percentage of combined carbon permits more ready remelting, hence less time for oxidation. Being more free from sand, less slag is formed in remelting, less fuel and flux are needed in the cupola, and cleaner castings result. The advantage of clean, uniform pigs for steel melting is unquestioned.

Composition and Appearance.—The external appearance of the solidified pigs serves to some extent as an indication of composition. Irons high in silicon generally give a full, rounded appearance to the top of the pigs. Conversely, irons low in silicon generally show hollowed tops and sharp, aggressive edges. A wrinkled, worm-eaten appearance to the top of the pigs generally accompanies a high sulphur content.

CHAPTER V.

ACTION WITHIN THE FURNACE.

Introductory.—The factors which compose the materials used in the blast furnace are four, namely, fuel, ore, flux and air. We have seen that of these four the first three are charged at the top of the furnace, while the last is forced in near the bottom. In consequence, we have in the furnace what may be described as two currents, traveling in opposite directions; a slow current of solids descending and a rapid current of gases ascending. The ascending current goes about as far in a second as the descending current goes in an hour. Although the former is gaseous, its weight per ton of product is about double that of the solid materials.

THE DESCENDING CURRENT OF SOLIDS.

When the bell is lowered and the charge slides into the furnace, we have a relatively cold, and often wet, body entering a heated atmosphere. The temperature at the top of the furnace is usually 400 to 600 degrees F., and heat is rapidly absorbed as the material passes downward. According to Sir Lowthian Bell, the temperature of the stock in an 80 foot furnace has risen to 1000 degrees F., when the stock has reached a depth of 10 to 12 feet, 1500 degrees F., at 20 feet., and 1800 degrees ·F. at 30 feet. According to Le Chatelier the temperature before the tuyeres exceeds 3500 degrees F., when all portions of the charge present exist in a state of complete fusion.

Fuel.—The effect of the heated gases upon each of the three kinds of substances in the charge as they journey through the furnace is very different. The fuel is least affected. It absorbs heat and a small percentage is dissolved by CO_2, but it suffers otherwise very little change until it arrives before the tuyeres. There it comes in contact with the blast, and its carbon is rapidly consumed, forming CO_2 which is immediately reduced to CO. This gas, mixed with the residual nitrogen of the blast, forms the

182

bulk of the upward current. The ash of the fuel continues downward and enters the slag.

Flux.—The change in the flux is of a very different nature from that of the fuel, and begins much sooner. It should be recalled that the flux is a calcic or magneso-calcic carbonate, having the formula, $CaCO_3$, or $Ca(Mg)$ CO_3 When the stone has attained a temperature of about 1100 degrees F., it begins to decompose and loses some of its CO_2, which joins the upward gaseous current, leaving as a residue a corresponding quantity of burned lime, CaO. Thus:

$$CaCO_3 = CaO + CO_2.$$

As the temperature of the stone increases, the decomposition is more rapid, and it is probable that by the time it has reached the bosh of the furnace, practically all of the CO_2 has been expelled and only the CaO remains.

$CaCO_3$ begins to decompose whenever its **vapor pressure** equals that of the superimposed atmosphere. Under reduced pressure, it will begin to decompose at about 1000 degrees F. In the ordinary atmosphere, however, the decomposition cannot become complete until the vapor pressure equals the atmospheric pressure, which requires a temperature of 1493 degrees F. As the pressure in the bosh of a modern furnace usually approximates $1\frac{1}{2}$ atmospheres, the final decomposition there of $CaCO_3$ cannot occur at temperatures much below 1600 degrees F.

CaO is infusible even at the highest temperature of the blast furnace hearth. However, by the time it has reached the hearth, it unites with the siliceous and aluminous gangue of the ore and ash of the fuel, and forms a fusible silicate or slag, commonly called "cinder," which drips down and accumulates in the crucible.

Ore.—The changes in the ore take place for the most part in the top of the furnace. It is usually the first member of the charge to be attacked. Unlike the stone, its first change is not due to heat alone, but to the reducing power of the gases, and is, therefore, chemical in its nature. The deoxidation begins at temperatures differing with the character of the ores. Easily reduced ores are affected at 400 degrees F., and the action becomes more rapid as the temperature rises. The change is usually about completed by the time the temperature of 1000 degrees F. is reached, which corresponds to a depth of less than 20 feet. Through the loss of its oxygen, the ore is changed to a finely divided sponge of metallic particles, which undergoes little change until it reaches the zone of fusion, where it is melted.

THE ASCENDING CURRENT OF GASES.

The air for combustion enters the furnace through the tuyeres in the form of a blast under pressure ranging in different furnaces usually from 5 to 20 pounds per square inch, and occasionally higher. It is sometimes used at atmospheric temperatures, when it is known as "cold blast," but generally it has a temperature between 800 degrees F. and 1400 degrees F, when it is known as "hot blast." The blast comes immediately into contact with the highly heated coke, the carbon of which unites with the oxygen of the blast, forming CO_2. Any moisture which may be present in the blast is broken up in the presence of carbon at this high temperature, forming CO and free hydrogen, thus:

$$C + H_2O \rightleftharpoons CO + H_2$$

The CO_2 formed by the combustion of the coke comes into immediate contact with incandescent particles of carbon, and is at once resolved into CO by the "carbon transfer," thus:

$$CO_2 + C = 2CO.$$

The following analyses by Van Vloten illustrate these interchanges·

	O.	CO$_2$	CO.	H.	N.
Middle of tuyere	13.0	6 0	0.6	0.75	80 50
Edge of tuyere	0 0	13.5	6.0	0.25	80.25
Between tuyeres	0.0	0.0	33 75	1 75	64.50

J. S. C I, 1893, p. 928.

From these analyses it appears that directly in front of a tuyere, free oxygen may still be found, and that it is associated with CO_2, but not with CO. At the edge of the tuyere the oxygen has all been consumed to CO_2 and some of the CO_2 has been already reduced by fresh carbon to the condition of CO. Between tuyeres, neither free oxygen nor CO_2 is found, but all of the carbon exists in the form of CO. As a result, the gaseous

current starts upon its upward journey with the following approximate composition by volume:

CO .... 35 per cent
II and Hydrocarbons. .... 1 per cent.
N ....64 per cent.

CO is an unsaturated compound with a strong tendency to
take up oxygen, thereby forming CO_2, thus:

$$CO + O = CO_2.$$

This action is strong enough to attract oxygen from combination with iron at elevated temperatures, although it is largely
neutralized by the presence of any considerable proportion of
the resulting CO_2

Action of CO.—The action of CO upon oxides of iron is
characterized by two decidedly marked phenomena, namely, the
removal of oxygen and the deposition of carbon. These two
reactions usually proceed simultaneously, but from the nature of
the changes it is evident that the reduction must occur first. The
action of CO on ferric oxide may be represented thus:

$$3CO + Fe_2O_3 = Fe_2 + 3CO_2.$$

In practice, however, this reaction is never complete, and a
considerable excess of CO is necessary to make it even approximately complete, as the presence of the resulting CO_2 tends to
undo the work of the CO. The reduced metallic iron is left in
the form of a finely divided sponge, which is keenly susceptible
to reoxidation. At elevated temperatures the CO_2, formed by the
above reaction, is capable of again giving up a part of its
oxygen to the spongy iron in accordance with the following
reaction:

$$Fe_x + yCO_2 = Fe_x O_y + yCO,$$

which acts more vigorously as the temperature rises. It is
probable that at all temperatures above 760 degrees F. the
reduction of ore by CO and the oxidation of the resulting sponge
by CO_2 can take place, and but for the law of " Mass action,"
a deadlock would exist in all but the very bottom of the furnace.
The great excess of CO as compared with CO_2 in all parts of
the furnace enables it to maintain activity, even at high temperatures, through its greater concentration.

The second phenomenon of the action of CO upon ores,

namely the deposition of carbon, is due to the power of the metallic sponge and the lower oxides of iron to split up CO in accordance with the following reaction:

$$2CO = CO_2 + C,$$

by which the separated carbon is deposited as a fine dust on the metallic sponge In this reaction the iron appears to act only as a "catalyser," since it does not enter into the reaction, but only facilitates it.   The quantity of carbon so deposited may amount to several times the volume of the ore.   As it is deposited in all the crevices and spaces of the ore, it exerts a very powerful disintegrating effect upon it   It may serve also to remove from the ore the last traces of oxygen left by the incomplete action of the CO.

From experiments in the laboratory of Sir Lowthian Bell at Clarence, England, it is evident that the activity of CO is affected not only by its purity, but also by the rate and pressure of the gas current, as shown by the following table:

	Oxygen extracted		Carbon deposited	
Kind of ore.	Slow current.	Fast current	Slow current	Fast current
Precipitated Fe_2O_3	49.30	80 60	79 70	335 40
Elba ore 16 90	16 90	18 20	3 80	4 90
Cleveland ore.............	37 30	50 70	12 60	22.30
Roasted spathic ore..... ....13 00	13 00	42 00	2 30	3.90

Age, 1897.

It is not likely that the velocity of the current increases the activity of the gas. The effect is attributable rather to the prompt removal of the neutralizing presence of the CO_2 formed by the two reactions Pressure, however, assists materially in the action as it brings the substances involved into firmer and more intimate contact.

E, 282.

It was shown by Laudig that the rates of reduction and carbon deposition are not dependent upon the state of division of the ore, but appear to be profoundly affected by the mineralogical classification He found the degree of susceptibility of the usual ores to be in the following order:

E, 269.

	Carbon deposited per cent of original ore	Carbon deposited per unit of iron
Mesaba ores............... . ..	21 61	0.3401
Brown hematites	12 94	0 2400
Red hematites ... .13 82	13 82	0 2206
Blue hematites	3 08	0 0501
Magnetites	0.10	0 0017

Action of H. —Free hydrogen also is a strong reducing agent. When passed over heated oxide of iron, it deoxidizes it even more readily than CO, in accordance with the reaction :

$$Fe_2O_3 + 3H_2 = Fe_2 + 3H_2O.$$

However, the fact that hydrogen occurs in gases escaping from the furnace seems to indicate that it does not perform reduction in the furnace. But while it may not reduce ore directly, it may assist in the reduction by diluting the CO_2 or even by decomposing it, producing CO thereby, thus :

$$H_2 + CO_2 = H_2O + CO,$$

which tends indirectly to faciliate reduction. The presence of hydrogen has another beneficial effect, in that it facilitates the decomposition of the limestone, as shown by the following experiments by Bell :

Temperature.	Atmos.	Time, minutes.	CO_2 removed.	CO_2 unde-composed.	CO_2 de-composed.	H_2O formed.
Bright red	Air	40	13 5	13 5	0	0
Bright red	CO_2	30	10 7	...	0	0
Bright red	H_2	30	40.5	21 4	19.1	7.44

The presence of hydrogen is due chiefly to the decomposition at the tuyeres of the moisture from the blast or other sources, thereby forming water gas, in accordance with the following reaction :

$$H_2O + C = H_2 + CO$$

The quantity present in the blast varies at different seasons of the year and different conditions of the weather. It usually ranges from $\frac{1}{2}$ to $1\frac{1}{2}$ per cent. of the gases by volume, the average for the year being about that yielded by 3 7 grains H_2O, per cubic foot of air, which amounts to 0.9 per cent. of the gases at the hearth and 0.8 per cent. at the downtake.

Action of N. —The nitrogen of the gases is practically inactive except for a slight tendency to unite with the hot carbon to form cyanogen, thus :

$$C + N = CN,$$

and to unite with hydrogen to form ammonia, thus :

$$N + H_3 = NH_3$$

Both cyanogen and ammonia act as deoxidizers, and therefore the gases start on their upward journey as a powerfully reducing organization

INTERACTION OF THE CURRENTS.

The objects which the gases encounter as they rise from the hearth are the incandescent coke, interspersed with drops of molten iron and slag. The melting zone of the furnace has its origin at the tuyeres, where the heat of combustion is liberated, and extends upward to a greater or less height, according to conditions. Usually, in a normally working furnace, it extends well up into the boshes. The top of the zone is marked by the melting of the reduced iron sponge. The molten iron then trickles down through the bed of incandescent coke, dissolving carbon as it goes, and passing through the layer of accumulated slag into the crucible. The melting point of the carburized iron rarely exceeds 2200 degrees F., but the fusion of the partly carburized sponge probably requires a temperature higher by several hundred degrees. A short distance below the fusing zone of the iron, the siliceous residue of the ore and part of the lime from the calcined limestone unite, become fluid and flow to the hearth. Slags, when once formed, can usually be remelted at temperatures of 2600 or 2700 degrees F., but owing to the lumpy condition and lack of intimate mixture of the ingredients, the temperature of slag formation is usually several hundred degrees higher. The ash of the fuel reaches the tuyere unchanged on account of its protecting coat of carbon, but as soon as the carbon is burned away the ash unites with the balance of the lime and joins the rapidly accumulating cinder.

Toward each of these substances, the gases in the hearth are neutral. The first change which comes to the gaseous current in its upward journey is the addition of CO_2, liberated by the decomposing limestone. This usually amounts theoretically to about 3 per cent. of the gases by volume, but is rapidly changed to CO in the presence of the hot coke, and, therefore, analysis rarely shows over 1½ per cent. in the zone of decomposition. Hence, by the time the gases have risen to a point 20 feet from the top, they show relatively little change in composition. From that point upward the greater part of the reduction of ore and the deposition of carbon takes place and the CO of the gases is rapidly converted into CO_2.

Since the carbon transfer cannot take place below 760 degrees

F., it follows that it is very desirable to keep the main evolution of CO_2 in the extreme top of the furnace, if possible. This fact puts a premium on easily reduced ores which give up the bulk of their oxygen at temperatures below that of the carbon transfer. Dense ores, which resist the action of the gases until they have descended to hotter parts of the furnace, always require a higher fuel consumption to compensate for the large loss of carbon due to the action, in the hotter parts of the furnace, of the CO_2 which has been evolved by the reduction.

We have, then, three sources of CO_2 in the furnace: (1) decomposition of limestone, $CaCO_3 = CaO + CO_2$; (2) reduction of ore, $Fe_2O_3 + 3CO = Fe_2 + 3CO_2$; (3) deposition of carbon, $2CO = C + CO_2$.

The first case differs from the other two in that the CO_2 evolved by the stone is an addition to the gas current and moreover is not a stable addition, since it occurs so low in the furnace that it is practically all changed to CO by the carbon transfer. In the other two cases the CO_2 is the result of chemical changes within the gases, and the changes are permanent.

This constantly increasing proportion of CO_2 tends to neutralize the reducing powers of the CO in two ways: by dilution, and by its tendency to give up part of its oxygen to the freshly reduced spongy iron. This latter tendency decreases with lowering temperature, and practically ceases at 760 degrees F. We can safely assume, therefore, that in the extreme top of the furnace only reducing conditions can exist.

Composition of Gases at Various Depths. —The change in composition of gases at various levels of the furnace is an indication of the reactions which take place at those levels:

Depth, feet.	Percentage by volume. CO. feet.	CO₂. feet.	Depth. feet.	Percentage by volume. CO. feet.	CO₂. feet.
Top	29.5	11.0	16½	34.1	2.2
4	29.5	10.5	20	35.1	0.7
8	27.0	8.0	39	35.0	1.1
10	32.0	7.0	52	35.2	1.5
12	33.0	7.0	65	35.9	0.5
14	31.0	6.5	70½	36.6	0.0
			Tuyeres	37.7	0.8

From inspection of the column of CO_2, the following cor-
roborative conclusions may fairly be drawn that CO_2 is formed
at the tuyeres, but that it is immediately changed to CO by the
presence of incandescent coke; that some undecomposed limestone reaches a depth of 65 feet, and CO_2 from that source is
evolved from that point to a point about 20 feet from the top;
that the carbon transfer ($CO_2 + C = 2CO$) takes place at all
points between the same levels; and that above 20 feet the
reduction of ore takes place with a consequent rapid increase
of CO_2.

Action of CO_2.—The oxidizing power of CO_2 is manifest at
all temperatures above 700 degrees. At 710 degrees, according to
Ackerman, it attacks carbon, and at 760 degrees, Bell found that
it oxidized the iron sponge. The latter reaction is partially offset, however, by the power of the deposited carbon to reduce
oxide of iron at all temperatures above 720 degrees F This
fact and the action of carbon in changing CO_2 to CO at all temperatures above 710 degrees prevents the decomposing limestone
from undoing the work which the CO has done in the top of the
furnace.

Action of Deposited Carbon.—Carbon which has been deposited from CO at low temperatures in the top of the furnace may be
carried downwards, dissolved by CO_2 and pass upward in the
form of CO again, only to be redeposited before escaping from
the furnace and brought once more into the region of the carbon
transfer. In this way an indefinite cycle may occur. The
activity of CO_2 in attacking deposited carbon, however, must
not be considered an unmitigated evil, since it tends to remove
what may become a serious obstruction to the passage of the
gases. The net result of this reaction is that we find very little
CO_2 in the gases at a temperature above 760 degrees F. On
the other hand, the action permits comparatively little of the
deposited carbon to reach the hearth. Carbon deposition ceases at
900 degrees F., and from that point to the hearth it is subject
to the continuous attacks of CO_2, and residual oxygen of the
ore, and may also supply some of the impregnated carbon of
the pig.

Summary.—These inter-reactions may be conveniently summarized as follows:

400 degrees F. and upward—CO reduces oxides of iron with the formation of CO_2, which is permanent.

$$Fe_2O_3 + 3CO = 2Fe + 3CO_2.$$

430 degrees F. to 900 degrees F.—CO is decomposed by the sponge and lower oxides, carbon is deposited and CO_2 formed, which is permanent.

$$2CO + Fe_x = Fe_x + C + CO_2.$$

710 degrees F. and upward—CO_2 attacks solid carbon, forming CO.

$$CO_2 + C = 2CO.$$

720 degrees F. and upward—Solid carbon reduces oxides of iron with formation of CO.

$$C + FeO = Fe + CO.$$

760 degrees F. and upward—CO_2 oxidizes metallic sponge, with the formation of CO.

$$CO_2 + Fe_x = Fe_xO + CO$$

1100 degrees F. and upward—$CaCO_3$ is decomposed with the evolution of CO_2, which is at once resolved into CO

$$CaCO_3 = CaO + CO_2.$$

From these facts, certain obvious conclusions may be drawn:

(1) CO can never wholly reduce Fe_2O_3 to the metallic state. The reduction is always accompanied by the evolution of CO_2, which partly neutralizes the action of the CO. For example, Bell found that Fe_2O_3 and iron sponge, acted upon simultaneously by pure CO for five hours at red heat, each contained at the end, 1 per cent. of the O necessary to form Fe_2O_3. The presence of the CO_2 from reduction in the one case and the carbon deposition in the other, was sufficient to maintain such a residue of oxygen.

(2) It is evident that for every mixture of $CO + CO_2$ at a given temperature, or for every temperature of the given mixture there is a definite point of equilibrium below which reduction cannot go, or beyond which oxidation cannot occur. Thus Bell found that when

$CO + CO_2$ acted on ore for several hours at bright red......28.65 % O remained.
When $CO + CO_2$ acted on spongy iron for several hours at
  bright red.. ....28.50 % O remained.
    Which shows equilibrium point for equal parts of CO and CO_2.
When $2CO + CO_2$ acted on ore for six hours at low red,......37.0 % O remained.
When $3CO + CO_2$ acted on same ore for six hours at low red..13.5 % O remained.
    Which shows that the point of equilibrium varied with the gas mixture

(3) When a point of equilibrium is reached, an increase of temperature or an increase of CO_2 will oxidize the product, while a decrease of either will have the opposite effect. Hence,

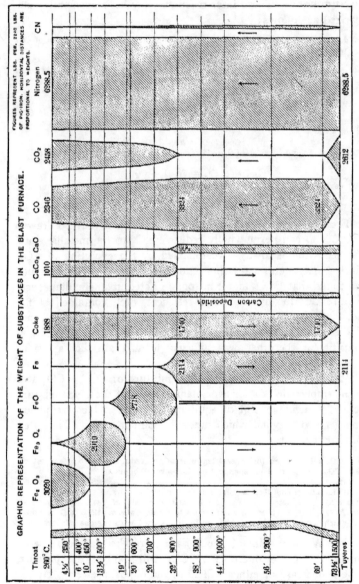

Graphic Representation of the Weight of Substances in the Blast Furnace.

since it is desirable from a point of fuel economy to have the ratio of CO to CO_2 as low as possible, the top temperature should be kept low to counteract the oxidizing effect.

(4) Since carbon is deposited at low temperatures, and its reducing effect is not diminished by high temperatures, we have a means of reducing the last traces of oxygen which are not removed by CO in the presence of CO_2.

The changes which take place in the materials at various depths and temperatures while passing through the furnace have been platted graphically by Dougherty as shown on page 192.

This plate is very instructive in that it enables the eye to help the brain grasp the various simultaneous changes which take place, but it is constructed from assumptions not entirely in accord with those usually accepted. The tuyere temperature given as 1500 degrees C. has generally been observed to be above 1900 degrees C. At a depth of 19 feet the ore is given as entirely in the state of FeO, which implies that only one-third of the oxygen has been removed at that depth, thus $Fe_2O_3 = 2FeO + O$, whereas Bell has shown that fully 85 per cent. is removed during the first ten or twelve feet. Moreover the diagram shows no metallic iron before the depth of 26 feet, while Bell found metal above 10 feet depth. Bell states also that carbon deposition ceases at red heat and owing to the action of CO_2 little or none reaches the bosh, but the diagram indicates that it persists even down to the tuyeres. The decomposition of limestone is here represented as sudden and complete at a depth of 32 feet, at which point the CO_2 column has its origin, whereas Bell's analyses showed the presence of CO_2 and undecomposed limestone at depths of 60 to 70 feet. Bell's analyses also show that cyanides disappear completely from the gases before they have risen 45 feet from the tuyeres.

THE CARBON RATIO.

A fair average composition by volume of the gases which escape from a blast furnace is as follows:

CO	CO_2	H.	CH_4.	N.
24.5	12.0	1.25	0.25	62.0

It was formerly believed that the ratio, $\dfrac{CO}{CO_2}$ could not fall below 2, and proper reducing effect in the furnace be maintained.

It has been shown repeatedly, however, that a furnace can be operated successfully for considerable periods with a ratio of 1.35 or less. This ratio of the unsaturated to the saturated carbon in the gases is of vital importance to the fuel economy. The heat liberated by the combustion of 1 pound of carbon, when burned to the state of complete oxidation, is 14,550 B. T. U., while that liberated by the unsaturated compound is only 4450 B. T. U per pound. Since over three times the heat is produced by the combustion to CO_2, it follows that the greater the proportion of CO_2 in the gases, the more efficiently each pound of fuel has done its duty, and the less fuel will be needed to keep up the furnace temperature. The importance of the carbon ratio to fuel consumption may be easily calculated:

Ratio CO to CO_2 1.85 = (0.65 × 4450) + (0.35 × 14,550) = 7985 B. T. U. per unit C.

Ratio CO to CO_2 2.33 = (0.70 × 4450) + (0.30 × 14,550) = 7480 B. T. U. per unit C.

Decrease per pound carbon in quantity of heat development = 505 = 6 per cent. C

which must be made up by burning additional fuel, which in turn, is less effective, on account of being less completely oxidized.

Quantity of CO and CO_2.—While the ratio of CO to CO_2 varies with the fuel consumption, it should be observed that the *quantity* of CO_2 per 100 pounds of pig is practically constant. This becomes self-evident when it is recalled that the CO_2 results entirely from the addition of oxygen in the reducing zone to the CO from the hearth. The total quantity of oxygen to be extracted from the ore per 100 pounds of pig varies within very narrow limits, and can never be very far from 43 to 44 pounds. Since only about 90 per cent. of the reduction is accomplished by the gases, it follows that about 39 pounds of oxygen per 100 pounds of pig will usually be taken up by the gases. This quantity of oxygen demands about 29 pounds of carbon, from which it appears that the quantity of carbon existing in the gases as CO_2 will never be far from 29 to 30 pounds for each 100 pounds of pig. Consequently, the ratio will depend solely upon the variation in CO. It is evident, therefore, that each rate of fuel consumption will be characterized by a different but tolerably definite carbon ratio. For a given set of conditions this ratio may be calculated approximately from the fuel consumed. The problem involves three factors the carbon in the fuel used, the carbon added by the limestone and the carbon withdrawn by carbon impregnation. The algebraical sum of these factors, less 29, gives the weight of carbon as CO.

Variations in Carbon Ratio.—If we assume that 5 7 pounds carbon is added per 100 pounds pig by the use of 50 pounds stone, and that 3 4 pounds carbon enters the pig, making a net addition of 2.3 pounds, then the carbon ratio for various fuel consumptions may be found as follows:

Fuel consumption per ton pig coke, 85 % fixed C, pounds	Pounds C from fuel per 100 pounds pig.	Net addition C of stone, less that impregnated.	Total C in gases.	C as CO_2	C as CO.	Carbon ratio.
1,700	64 31	2 30	66 81	29	37 81	1 31
1,800	68 34	2 30	70 64	29	41 64	1.44
1,900	72 08	2 30	74 38	29	45 38	1.57
2,000	75 80	2 30	78 10	29	49 10	1 70
2,100	79 69	2 30	81 99	29	52 99	1.83
2,200	83 47	2 30	85 77	29	56 77	1 96
2,300	87 30	2 30	89 60	29	60 60	2.09
2,400	91 07	2 30	93 37	29	64 37	2 22
2,500	94 86	2 30	97 16	29	68 16	2.35
2,600	98 60	2 30	100 90	29	71 90	2.48
2,700	102.42	2.30	104 72	29	75 72	2 61
2,800	106 25	2 30	108 55	29	79 55	2 74
2,900	110 07	2 30	112 37	29	83 37	2 87
3,000	113 81	2 30	116 11	29	87.11	3 00

CHEMICAL REACTIONS.

Of the chemical reactions which take place in the furnace, seven may be distinguished as being persistent and authentic Of the seven, three may be grouped together as being heat-giving, or exothermic, and four as heat-absorbing, or endothermic.

Exothermic Reactions.—The exothermic reactions are as follows:

(1) Oxidation of carbon to carbon monoxide by the free oxygen of the blast and the residual fixed oxygen of the ore, thus, $2C + O_2 = 2CO$, by which there is developed 4450 B. T. U. per pound of carbon.

(2) Oxidation of part of the carbonic monoxide to carbonic dioxide through reduction of the iron of the ore, thus, $Fe_4O_3 +$

$3CO = 2Fe + 3CO_2$, by which there is developed 10,100 B. T. U. per pound of carbon.

(3) Oxidation of another portion of the carbonic monoxide to carbonic dioxide through the deposition of carbon, thus, $2CO = C + CO_2$, by which there is developed $10,100 - 4450 = 5650$ B. T. U. for each pound of carbon so oxidized.

Endothermic Reactions.—The endothermic reactions are as follows:

(1) The decomposition of limestone into lime and carbon dioxide, thus. $CaCO_3 = CaO + CO_2$, which absorbs 812.7 B. T. U. per pound of stone or $\dfrac{812.7 \times 100}{12} = 6772\,5$ B. T. U. per pound of carbon in the stone.

(2) The reduction of carbonic dioxide by carbon, thus: $CO_2 + C = 2CO$, thereby absorbing $14,550 - (2 \times 4450) = 5650$ B. T. U for each pound of carbon reduced.

(3) Reduction of carbonic dioxide by metallic iron or the lower oxides of iron, thus $CO_2 + xFe = Fe_xO + CO$, absorbing $14,550 - (6300 + 4450) = 3800$ B. T. U per pound of carbon involved.

(4) Decomposition of moisture into its elements, thus: $C + H_2O = CO + H_2$, thereby absorbing $6890 - 2965 = 3925$ B. T. U. per pound water, or 5885 B T. U. per pound of carbon involved.

Reduction.—The reduction of oxides of iron by carbon may take one of three forms.

$$(1)\quad 3C + Fe_2O_3 = 2Fe + 3CO,$$

which liberates 4450 B. T. U. per pound carbon, or $4450 \times \dfrac{12}{16} = 3337$ B. T. U. per pound oxygen removed, and 7335 absorbed, a loss of 3997 B. T. U.

$$(2)\quad 3C + 2Fe_2O_3 = 4Fe + 3CO_2,$$

which liberates 14,550 B. T. U. per pound carbon, or $14,550 \times \dfrac{12}{32} = 5455$ B. T. U. per pound oxygen removed and 7335 absorbed, a loss of 1880 B T. U.

$$(3)\quad 3CO + Fe_2O_3 = 2Fe + 3CO_2,$$

which liberates 4330 B. T. U. per pound CO, or $4330 \times \dfrac{28}{16}$

= 7577 B. T. U. per pound oxygen removed and 7335 absorbed, a gain of 242 B. T. U.

The last reaction gives the most heat per unit of oxygen, and hence is most desirable from the standpoint of heat development. Moreover, it is exothermic, and when once begun is capable of running without further additions of heat.

Zones of Reduction.—As we have seen, the action of CO in the top of the furnace removes the bulk of the oxygen from the ore before it has descended far. Bell found in his researches that from certain ores, 76 per cent. of the oxygen had been removed at a depth of 10 feet and 18 per cent. still remained at 70 feet. With the Lake ores in this country, however, there is good reason for believing that fully 90 per cent. of the oxygen is removed in the upper zone of reduction, and that the ore undergoes little change beyond heat absorption between the depth of 20 feet and the zone of fusion. In the lower part of the furnace, however, there is a second zone of reduction, where the metalloids are deoxidized and the last traces of oxygen are removed from the iron.

Chem.:
p. 42

HEAT DEVELOPMENT IN THE FURNACE.

The heat developed in the furnace may be ascertained with tolerable accuracy if the quantity and analysis of the escaping gases are known. It will evidently be the weight of carbon burned to CO_2, multiplied by 14,550, plus the weight of carbon burned to CO, multiplied by 4450. It is only necessary to ascertain the quantity of these gases per unit of iron. This can be easily done if the quantity of fuel and flux and the composition of the resulting iron are known.

Let us assume an iron of the following composition:

	Per cent.		Per cent
C	3.40	S	0 05
Si	1.50	Fe	94 00
Mn	0 50		
P	0 50		99.95

to be smelted with 1 pound of coke and ½ pound of stone for each pound of pig, and that the slag weighs about 55 per cent. of the iron. Let us assume, also, that the coke contains 85 per cent. fixed carbon and 0.5 per cent. moisture and that 90 per cent. of the oxygen in the ore is removed by the gases in the upper part

of the furnace, leaving the other 10 per cent, including that combined with the metalloids, to be removed by solid carbon. Let
us assume, also, as a fair average, that it takes 1.7 tons of ore
to make a ton of pig containing 94 per cent. metallic iron, and
that the ore contains 3 per cent. combined water and 7 per cent.
moisture, giving 5.1 and 11.9 pounds, respectively, of water per
100 pounds of pig, and that the limestone and coke each contain
0.5 per cent. moisture, yielding 0.25 pounds and 0.5 pounds more
water.

In order to ascertain the quantity of heat developed in the
furnace, it is necessary to find, first, how much carbon is burned
with heating effect, then how much is changed from CO to CO_2
by reduction in the furnace top. The rest will be the CO.

The oxygen to be removed from:

	Pounds
94.0 pounds Fe =	.40.30
3.4 pounds C =	0.09
1.5 pounds Si =	1.71
0.5 pounds Mn =	0.29
0.5 pounds P =	0.65
0.05 pounds S =	0.05
99.95 pounds pig iron =	43.00
Total oxygen to be removed =	.43.00
90 per cent removed by CO =	.38.70
Leaving to be removed by solid C	4.30
100 pounds coke at 85 per cent. fixed carbon =	85.0
C impregnated in pig	3.4 pounds.
C stolen by CO_2 from 50 pounds stone	5.7 pounds.
	9.1
C left to be burned	.75.9
C burned by 4.3 pounds fixed O	3.23
C left to be burned by the blast.	.72.67

The CO_2 in the gases is all formed by the action of CO upon the ore,
from which it takes up 90 per cent. of the oxygen, or 38.7 pounds

The quantity of carbon which can take up 38.7 pounds O is $\dfrac{12}{16} \times 38.7 = C$

as CO_2 =	.29.00
C burned to CO by the blast and remaining as CO =	43.67 pounds.
C burned to CO by fixed oxygen =	3.23 pounds.
C as CO =	.46.90
Total carbon consumed as above	.75.90

The heat development of the furnace may now be shown as
follows:

B. T. U.

29 0 pounds C burned to CO_2, releasing 14,550 B T. U = ...421,950
46 9 pounds C burned to CO, releasing 4,450 B T U = ...208,705

Total heat development by fuel = ...630,655
Less heat absorbed by reduction of 5 7 pounds C from
   CO_2 to CO at . . . . .10,100 B. T. U.
And oxidation simultaneously of 5 7 pounds C from C to
   CO at .... 4,450 B T. U.

Leaving a net loss of... .... 5,650 B. T U = 32,205

Whence the net heat development in the furnace =..... 598,450
75 9 pounds C burned to CO_2 develops 1,104,345

which shows that the heat development in the furnace is 54 2 per cent. of the theoretical, or 7885 B. T. U. per pound carbon, of which 66 9 per cent. is due to CO_2, and 33.1 per cent to CO.

A further analysis of these figures shows that the heat developed in the hearth consists of:

B T U.

75.9 pounds C burned to CO at 4 450 B. T U. per pound = ...337,755
while that developed in the furnace top consists of 29 0 pounds C as CO
   burned to CO_2, at 10,100 B T. U per pound = ...292,900

                                     630,655
Less the heat absorbed by carbon transfer 32,205

Net heat developed as above............................ 598,450

of which 53 5 per cent is developed in the hearth, and 46.5 per cent in the top of the furnace.

Carbon Ratio.—In order to find the carbon ratio existing in the final gases, it is necessary to take into account the carbon from the stone, and that stolen by it, as follows:

C burned to CO and remaining CO 46 9
C existing as CO from the CO_2 of stone 5 7
C existing as CO, stolen by CO_2 of stone 5 7

                                         58 3

$$\text{The carbon ratio } \frac{CO}{CO_2} = \frac{58\ 3}{29\ 0} = 2.01$$

Heat in the Blast.—In addition to the heat developed by combustion, in the furnace, there is another source of heat to the hearth, namely the heat in the blast. The quantity so introduced often amounts to one-fifth of the total heat development in the furnace and one-third of that developed in the hearth. It may be determined by multiplying the weight of the air by its specific heat. The weight of the blast needed by the fuel per 100 pounds of pig may be found as follows:

	Pounds.
O_2 needed to form CO with 72 07 pounds carbon =	96.89
O_2 furnished by moisture in blast (annual average, 3 7 grains H_2O per cubic feet air) =	2.50
O_2 furnished by dry air =	94.39
N_2 accompanying O_2 of dry air =	310 55
H_2 accompanying 2 50 pounds O_2 in moisture =	0.31
Total weight of moist blast =	407.75

which at 60 degrees F. equals 5300 cubic feet per 100 pounds of pig. Owing to clearance and losses, this usually requires 5500 cubic feet piston displacement.

The heat brought in by the blast may be determined by means of the formula:

$$\text{Heat} = [0\,2335\,(t - t') + 0.0000208\,(t^2 - t'^2)] \times \text{weight of air.}$$

Between 60 and 1200 degrees F., the available heat in the blast per 100 pounds pig is as follows:

$$407.75 \text{ pounds } [0.2335 \times 1140 + 0.0000208 \times 1,436,400]$$
$$= 120,720 \text{ B. T. U.}$$

Based on the total heat developed in the furnace, this equals:

$$\frac{120,720 \times 100}{598,450} = 20\,2 \text{ per cent}$$

Based upon the heat developed at the tuyeres, it equals:

$$\frac{120,720 \times 100}{72.67 \times 4450} = 37.3 \text{ per cent.}$$

Temperature of the Hearth.—The approximate temperature of the hearth may be calculated from the total heat present at the tuyeres, if the weights and specific heats of the products of combustion are known. The weight of gases which results from the combustion of 72 67 pounds carbon to the condition of CO through the agency of moist air, is as follows:

CO	109 56 pounds
N_2	310 55 pounds
H_2	0 31 pounds
	480 42 pounds.

which is equal to 6.61 pounds per pound of carbon burned by the blast. The specific heats of N_2 and CO are the same, and that of the small quantity of H_2 may be neglected. The total quantity of heat present at the tuyeres is manifestly that developed by the combustion of the carbon, plus that brought in by the materials of combustion, thus:

B. T. U.

Heat developed by combustion per pound carbon = ... 4,450

Heat brought in by blast per pound carbon $\dfrac{120,720}{72.67}$ = ... 1,661

Heat brought in by C per pound carbon =..................... 0.5t — 216

Total heat present in the hearth =................. . . .. .0.5t + 5,895
Deducting that rendered latent in decomposing 2.81 pounds moisture
$\dfrac{2.81 \times 5,750}{72.67}$ =. 222

Net heat present in hearth =. .... .0.5t + 5,673
Substituting these values in the formula $(0.2405t + 0.00002143t^2$
$=$ total heat) gives 6.61 $(0.2405t + 0.00002143t^2) = 0.5t + 5,673$,
whence $1.5897t + 0.00014165t^2 = 0.5t + 3,673$ and, $0.00014165t^2$
$+ 1.0897t = 5,673$.
Completing the square and solving for t gives,
$$0.0119t = 42.3$$
$$t = 3,555 \text{ degrees F.}$$

This calculation, however, takes no note of the heat extracted from the hearth walls by radiation and cooling water. This action, which is exceedingly difficult to measure quantitatively, reduces the total quantity of heat present, and would make the result somewhat lower. The temperature observed by Le Chatelier was 3506 degrees F.

From a critical survey of the above calculation, it is evident that the weight of the products of combustion is always directly proportional to the weight of carbon, and will always be about 6.6 pounds per pound of carbon. For a given set of conditions, therefore, the temperature of the hearth is in equilibrium, no matter how much fuel is burned. The temperature will always be that of combustion under the given conditions. If, however, the conditions be changed, the equilibrium will be destroyed and a new temperature of combustion will result. For example, if the temperature of the blast be increased, or its percentage of moisture be decreased or coke at a higher temperature be precipitated into the hearth, a higher hearth temperature will follow. The maximum temperature is usually found three or four feet above the tuyere zone.

It should be observed here that the above temperatures mark the thermal state of the gases and not that of the products of fusion. Since the latter derive their heat from the former there must necessarily be a considerable difference of temperature in order that there may be appreciable interchange of heat during their brief contact as they travel in opposite directions. It is probable, therefore, that there is a constant difference of at least 300 to 400 degrees F. between the two currents at any given level of the furnace.

Zone of Fusion.—Since no smelting reactions take place at the tuyeres, the temperature at that place is of secondary importance. The really important point in the heat zone is the zone of fusion, as it is there that the products of the operation are subjected to temperatures which give them their distinctive characteristics Since the products of the zone of fusion melt at quite different temperatures, and therefore the respective points at which the meltings occur are necessarily separated by an appreciable distance, the zone of fusion must be looked upon as a wide band, or, better still, as two distinct zones, one above the other. The locating of these two zones in the furnace is not a simple matter, as they are subject to so many conditions and may fluctuate even within comparatively brief periods. The factors which contribute most to this uncertainty, aside from the quantity of heat developed by combustion, are: the degree of expansion of the gases, the latent heat of fusion of the products, the increase of specific heats of the products. The last two are accurately known quantities, but the first evidently depends upon the bosh angle and the position of the zone of fusion. If it is assumed, for example, that the bosh is low, and its area does not greatly exceed twice the hearth area, it might be presumed that by the time the gases have reached the slag-melting zone, they have lost, through cooling and increased space, about one-quarter of their expansive force. As we shall see later, the heat that would be absorbed during free expansion per 100 pounds of pig is about 80,000 B. T. U. Hence, under this supposition, 20,000 B. T. U. would be absorbed, or 275 B. T. U. per pound of carbon burned. The latent heat of fusion of slag per pound is about 180 B T. U. Assuming 55 pounds of slag per 100 pounds of iron, we will have: $\dfrac{180 \times 0.55}{0.7267}$ = 136 B. T. U. per pound of carbon burned. The specific heats of CaO and SiO$_2$ average about o.18 + o 00004t, which at 3000 degrees equals about o.30 B. T. U. per pound. The heat capacity of molten iron is o 22 B. T. U. per pound, and of carbon it is o.5 — $\frac{216}{\cdot t}$ B. T. U per pound. Allowing for the heat absorbed in raising the temperature of these substances 400 degrees, and that rendered latent by expansion of gas and fusion of slag, the formula for the temperature would be modified to read as follows:

$$6\ 61\ (0\ 2405t + 0.00002143t^2) + (1.367 \times 0.22 \times 400) + (0\ 76 \times 0\ 30 \times 400)$$
$$+ \left(0.5 - \frac{216}{3,000} \right) \times 400 = 0\ 5t + 5,673 - (275 + 136),$$

whence
$$0\ 00014165t^2 + 1.5897t + 120 + 91 + 171 = 0.5t + 5,673 - 411.$$

Transposing gives
$$0.00014165t^2 + 1.0897t = 4,880,$$

completing the square,
$$0.00014165t^2 + 1.0897t + 2,097 = 6,977.$$

Taking square root,
$$0.0119t + 45.8 = 83.5.$$
$$0\ 0119t = 37.7$$
$$t = 3,170 \text{ degrees F.}$$

which is the temperature of the gases at the level of slag fusion. Deducting about 350 degrees, for difference of temperature, gives 2820 degrees F. as the melting point of the slag.

By carrying the same reasoning a step farther we may investigate the zone of fusion of the iron. Assuming that the gases have lost another quarter of their expansive force, the additional latent heat of expansion performed of carbon would be 275 B. T. U. The latent heat of fusion of iron is 126 B. T. U. per pound, hence per pound of carbon it will be $\frac{126}{0.7267}$ = 173 B. T. U. Deducting these, together with the heat necessary to raise the temperature of the molten iron, the slag forming materials and the coke another 400 degrees, we have this formula: .

$$6\ 61\ (0.2405t + 0\ 00002143t^2) + (1\ 367 \times 0.22 \times 800) + (0.76 \times 0.30 \times 800)$$
$$+ \left(0\ 5 - \frac{216}{3,000} \right)\ 800 = 0\ 5t + 5,673 - (411 + 488)$$

Which equals
$$0\ 00014165t^2 + 1\ 5897t + 240 + 182 + 272 = 0\ 5t + 5,673 - 899$$

Transposing gives
$$0\ 00014165t^2 + 1.0897t = 4,080$$

Completing square,
$$0.00014165t^2 + 1.0897t + 2,097 = 6,177.$$
$$0.0119t + 45\ 8 = 78.6.$$
$$0\ 0119t = 32.8.$$
$$t = 2,755 \text{ degrees F.}$$

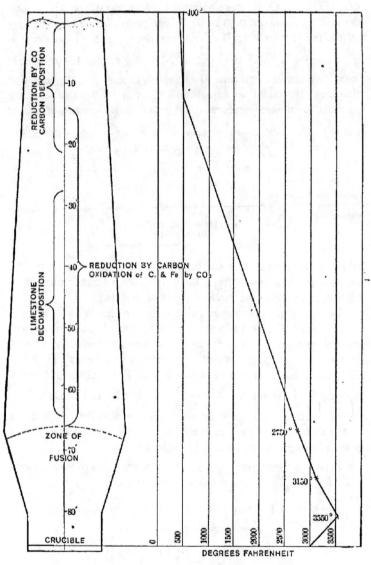

Temperature Gradient of Blast Furnace.

which is the temperature of the gases at the iron melting zone. Deducting 350 degrees for.the difference of current temperatures gives 2400 degrees, which is about the melting point of semi-carburized iron.

The temperature gradient of the furnace from hearth to top may be represented by the accompanying diagram, which is a modification of Allen's.

Heat Evolved by Reduction.—Between the zone of fusion of the iron and the zone of gaseous reduction very little heat is rendered latent beyond that absorbed in setting free CO_2 of the flux. Through this long distance the gases lose heat mainly through conduction to the materials of the charge, and therefore it has been termed the **zone of heat interception.** When the zone of reduction is reached, the gases meet a slight evolution of heat. The amount of heat evolved at this point is the excess of heat developed by the oxidation of CO to CO_2, less that absorbed by the reduction of Fe_2O_3 to Fe_2, in accordance with the reaction,

$$Fe_2O_3 + 3CO = Fe_2 + 3CO_2.$$

B T. U

29 parts C as CO oxidized to CO_2, developing 10,100 B. T. U = ...292,900
90 2 parts Fe reduced, absorbing 3,143 5 B. T. U. = ... 283,615

Heat liberated per 100 pounds pig = ... 9,285

Composition of Final Gases.—As the gases pass upward from the hearth, they make the following additions before escaping from the furnace:

43.00 pounds O from the ore.
3 23 pounds C burned by fixed O.
26.77 pounds CO due to carbon transfer
17.75 pounds moisture and combined H_2O from ore.
90.75 pounds.

Adding these to the gases at the tuyeres gives the composition of the escaping gases, as follows:

CO_2 .... .106.40 pounds.
CO .... 136.16 pounds
N_2 ....310.55 pounds
H_2 .... 0.31 pounds
Steam .... 17.75 pounds.

Total weight of escaping gases per 100 pounds pig........ ......571.17 pounds.

Temperature Due to Reduction.—If we assume that the materials and the products of combustion absorb all of this heat, it is a simple matter to calculate the resulting rise of temperature, as follows:

Ore, 1.7 $(0.1456t + 0.0000188t^2)$ =.......... ... . $.0.2475t + 0.0003196t^2$
Coke, 1.0 $(0.1567t + 0.00036t^2)$ = ... $0.1567t + 0.0003600t^2$
Stone, 0.5 $(0.2085t)$ =.......... .. .. $0.1042t$
CO_2, 1.064 $(0.187t + 0.000111t^2)$ =...... ....$0.1990t + 0.0001181t^2$
$N_2 + CO$, 4.467 $(0.2405t + 0.0000021t^2)$ = ...$1.0743t + 0.0000957t^2$
H_2O, 0.177 $(0.42t + 0.000185t^2)$ =.................... $0.0743t + 0.0000327t^2$

$$1.8560t + 0.0009261t^2$$

$$1.8560t + 0.0009261t^2 = 92.85.$$

Completing the square gives

$$0.0009261t^2 + 1.856t + 930.5 = 1,023.35.$$

Taking the square root,

$$0.03043t + 30.5 = 32.0$$

Whence, $\qquad\qquad\qquad t = 49$ degrees F.

However, there is not sufficient time for the solid products to absorb their full capacity of heat, and, therefore, the temperature of the gases is always considerably higher.

Heat Lost in Waste Gases.—The heat in the escaping gases at 450 degrees F. may be calculated by means of the formula for difference of temperatures, as follows:

B. T. U.
CO_2, 106.40 $(0.187 \times 390 + 0.000111 \times 198,900)$ =.. 10,108
$\left.\begin{array}{l} CO, 136.16 \\ N_2, 310.55 \end{array}\right\}$446.71 $(0.2405 \times 390 + 0.00002143 \times 198,900)$ =.·· 43,803
H_2, 0.31.$(3.367 \times 390 + 0.0003 \times 198,900)$ = ... 425
Steam, 17.75 $(0.42 \times 390 + 0.000185 \times 198,900)$ =.................. 3,560

$\qquad\qquad\qquad\qquad\qquad\qquad\qquad\qquad\qquad\qquad\qquad$ 57,896

which is the amount of heat that would be given up if the temperature were brought down to 60 degrees F. This heat is equivalent to 15 pounds of coke burned at the tuyeres per 100 pounds pig or 336 pounds per ton.

HEAT REQUIREMENT.

The heat requirements of a furnace making a given grade of pig on a given consumption of fuel and flux would be constant, were it not for changes in the weather, which introduce varying amounts of moisture into the furnace by means of the blast and materials of the charge. The *constant* heat requirement for manufacturing 100 pounds of pig under the present supposition may be stated as follows:

B. T. U.   B. T. U.

Reduction of 94.0 pounds Fe from Fe_2O_3 @................. 3,143.5 = 295,500
Reduction of 1 5 pounds Si from SiO_2 @.. ....11,571.5 = 17,357
Reduction of 0 5 pounds Mn from MnO_2 @................. 4,100.7 = 2,050
Reduction of 0 5 pounds P from P_2O_5 @ ....10,605.6 = 5,303
Reduction of 0.05 pounds S from SO_2 @............. .. 3 896 0 = 195
Reduction of 0 25 pounds Mn from MnO_2 to MnO @.......... 2,975.0 = 744
Decomposition of 5.1 pounds H_2O of hydration of ores @..... 1,300 0 = 6,630
Decomposition of 47 5 pounds $CaCO_3$ (from 50 pounds stone) @ 812 7 = 38,603
Reduction of 5 7 pounds C as CO_2 to CO @ .. .. .10,100 0 = 57,570
Reduction of 3 4 pounds C as CO to C @ .. .. 4,450 0 = 15,130
Fusion of 100.0 pounds pig iron @ .... 666 0 = 66,600
Fusion of 55 0 pounds slag @ . .. . 1,000.0 = 55,000

Total constant requirement 560,682

The *variable* heat requirements on the same
basis may be stated as follows:

Evaporation of 0 50 pounds moisture in coke @ .....1,120 0 = 560
Evaporation of 0 25 pounds moisture in stone @......1,120.0 = 280
Evaporation of 11 90 pounds moisture in ore @ ...... 1,120 0 = 13,328
Decomposition of 2.81 pounds moisture in blast @.. ..5,750.0 = 16,157
Carried off in 571.17 pounds waste gases @ 450 de-
grees F. .... = 57,896

Total variable requirement.... 88,221

Total requirement of 100 pounds pig, under above conditions 648,903

Unavoidable Losses of Heat.—By comparing the heat development and the requirements of the furnace per 100 pounds of pig, we have the following:

B T U.

Net heat developed in furnace by fuel 598,450
Heat brought in by blast................. 120,720

Total heat available 719,170
Total heat accounted for.. 648,903

Leaving unaccounted for............................... . 70,267

which amounts to about 9 77 per cent of that available, and which is irrecoverably lost through radiation from the furnace stack and conduction through the foundations and cooling water.

Heat Intercepted by Descending Materials.—The quantity of heat usefully intercepted by the column of descending materials will evidently consist of the total heat available in the hearth, less that which escapes through radiation, conduction, etc, plus that which is carried away in the waste gases. The amount of heat so returned to the hearth per 100 pounds pig may be calculated as

	B. T. U
Heat carried away in waste gases	57,896
Heat lost by radiation and conduction	76,867
Total heat lost to the furnace	134,763
Deduct heat developed in furnace by reduction	9,285
Net loss of heat which was available in the hearth	125,478
Total heat developed in hearth (75.9 pounds C to CO)	337,755
Heat brought in by blast	120,720
Total heat present in the hearth	458,475
Absorbed in decomposing 2 31 pounds moisture	16,157
Net amount of heat available in hearth	442,318

442,318 — 125,478 = 316,840 B. T. U., which represents the available heat intercepted by the descending materials. Figured on the basis of the total heat developed in the hearth this equals:

$$\frac{316,840 \times 100}{458,475} = \dots \dots 69\ 1\ \text{per cent}.$$

The heat utilized in decomposing the moisture of the blast equals,

$$\frac{16,157 \times 100}{458,475} = \dots \dots 3\ 5\ \text{per cent}.$$

The heat irrecoverably lost to the furnace equals,

$$\frac{125,478 \times 100}{458,475} = \dots \dots 27.4\ \text{per cent}.$$

This shows the blast furnace to have a thermal efficiency of 72½ per cent.

Effect of Heated Blast.—When the heated blast was first introduced it was found that it was accompanied by a saving of fuel out of proportion to the quantity of fuel necessary to heat it. This was not understood until it was shown that the combustion in the furnace hearth developed only 30 per cent. of the heat of the fuel. It then became clear that the blast, when heated outside of the furnace where complete combustion of fuel can take place, requires only about one-third of the fuel that is required inside the furnace. In additon to the extra fuel needed to heat the blast inside the furnace, there is an increased quantity of blast needed to burn the extra fuel to be heated, and a correspondingly increased loss of heat carried away by the increased quantity of waste gases. The quantity of coke burned to CO which would produce the 120,720 B. T. U. in the blast when heated to 1200 degrees F. is $\frac{120,720}{4450} \div 85 = 32$ pounds for each 100 pounds of iron.

When we take into account the increased quantity of blast needed to consume this fuel, which must also be heated, and, in addition, the losses of heat due to the increased quantity of gases, we can see at a glance that the fuel consumption of a cold blast furnace may easily exceed that of a hot blast furnace by 50 per cent.

Fuel Value of Hot Blast.—The value of increase in blast temperature in terms of coke illustrates the advantages of hot blast. If 407.75 pounds be the weight of blast needed per 100 pounds of pig, then the quantity of heat brought into the furnace for 100 degrees F. of blast temperature above 1200 degrees F. may be found by the formula .2335 $(t-t')$ + .0000208 $(t^2-t'^2)$ = heat

407 75 (0.2335 × 100 + 0000208 × 250,000) = 11,640 B. T. U., $\frac{11,640}{4450}$ ÷ 85 = 3 pounds coke, burned in the hearth per 100 pounds of pig produced, which is a saving of about 70 pounds coke per ton of pig, or 3 per cent. of fuel. A decrease of 3 per cent of fuel means also a decrease of 3 per cent. of blast, so that the quantity of heat actually brought in at that temperature would be somewhat less than indicated.

Minimum of Fuel.—If the duty of the fuel were simply to furnish a reducing agent for the ore, it would be sufficient to furnish enough carbon to remove the oxygen. Assuming that the reduction could be performed by solid carbon with the formation of CO_2, it would be necessary to use only $\frac{100 \times 36}{4 \times 56}$ = 16 05 pounds C, instead of 100 pounds in order to make 100 pounds of iron. But we have seen that a large excess of carbon in the gases is necessary to perform the reduction. Moreover, the carbon does not burn to the condition of CO_2 in the furnace but to CO, thereby giving out only about 30 per cent. of its available heat. This would appear to be a sad waste of calorific energy, were it not for the fact that CO is indispensable in preparing the ore for fusion, and moreover, however much sensible heat is intercepted, there is external work to be done by which this undeveloped energy may still be utilized. Since a portion of the fuel energy is released in the zone of combustion and a still further quantity in the zone of reduction, it is evident that there are two zones of heat development, of which the greater is in the lower furnace and the lesser in the upper.

Limits of Blast Temperature.—Since the use of heated blast is so effective in cutting down the need of fuel in the furnace, it might be argued that if the blast could be heated sufficiently, the iron could be made by hot air alone. Theoretically, the fuel could be replaced by the hot blast down to the 16 05 pounds of carbon per 100 pounds of pig that we found necessary to perform the reduction, were it not for the physical difficulty of putting into the blast the required quantity of heat. As the quantity of fuel decreased, a constantly less quantity of blast would be needed, and therefore each particle of the air would have to be heated to a correspondingly higher temperature. Assuming the heat requirement of the furnace to be as found above, 458,475 B. T. U. per 100 pounds of pig, of which 337,755 is furnished by the combustion of 76 pounds C to CO, and 120,720 by 407 75 pounds blast, heated to 1200 degrees F., then the necessary blast temperature for each decrease of fuel would be as follows:

Fuel per ton pig	Pounds C per 100 pounds pig.	Heat require-ment.	Heat of combustion.	Heat needed in blast.	Wt. air to burn C.	Tem-perature necessary, degrees F.
2,240	76	458,475	337,755	120,720	407.75	1,200
2,000	67	458,475	298,150	160,325	350.80	1,650
1,800	60	458,475	267,000	191,475	322.20	2,125
1,600	54	458 475	240,300	218,175	290 00	2,600

Effect of Cold Blast.—The gases which are given off by a cold blast furnace do not differ materially from those from a hot blast furnace, except that the ratio $\frac{CO}{CO_2}$ is higher, as a result of the higher fuel requirement. The presence of more CO with its attendant nitrogen gives a greater volume of gases, which must, in consequence of their greater volume, pass more rapidly through the cold stock, thereby being less thoroughly cooled. The smaller the quantity of gases, the more slowly they need to move to make place for that following, and consequently the more time they have to give up their sensible heat to the surrounding materials.

Volume of the Blast.—We have found that under the above

conditions, 407.75 pounds of air enters the furnace for each 100 pounds of pig, which corresponds to about 5300 cubic feet at 60 degrees F. In being heated to 1200 degrees F., the air expands enormously; its volume at 32 degrees F., being added for each rise of 490.5 degrees F. The old volume will then bear to the new the same ratio as their respective absolute temperatures. This relation may be expressed as follows:

$$1 : 60 + 458.5 = x : 1200 + 458.5,$$

by which it appears that the new volume, x, amounts to about 3.2 cubic feet for each cubic foot at 60 degrees F. However, as blast pressure in modern furnaces usually reaches 14 to 15 pounds per square inch, or about one atmosphere, it follows that the actual volume of the blast as it enters the furnace will be about 1.6 times the volume of the air at 60 degrees F.

The piston displacement required to furnish 407.75 pounds of blast varies considerably at different seasons of the year. For example, when the air has a temperature of 30 degrees F., each pound of it occupies 12.35 cubic feet of space, and hence, each 100 pounds of pig requires,

$$407.75 \times 12.35 = 5036 \text{ cubic feet.}$$

At 90 degrees, each pound of air has a volume of 14.22 cubic feet, and hence

$$407.75 \times 14.22 = 5798 \text{ cubic feet,}$$

which are required for the same purpose. This difference of 762 cubic feet represents an increase of over 15 per cent. in volume, and for a blowing cylinder 84 x 60 inches, requires 45 additional engine revolutions per ton of pig iron made

It is customary to measure the quantity of blast driven into the hearth by means of the piston displacement of the engine. This is a convenient method, but it is only approximate. After clearance space is deducted, it measures with tolerable accuracy the quantity of blast that leaves the cylinder when the valves are tight, but the quantity of air which enters the furnace will be lessened by just the amount of leakage between the engine and the hearth. Good connections should never require over 55 cubic feet piston displacement per pound of coke. Since the quantity of air needed by the furnace is absolutely dependent on the carbon burned, it should always be reckoned in terms of fuel. Any attempt to report the air consumption in terms of iron is always indefinite, unless the fuel consumption is known, since the quantity of air per pound of pig will vary according to the fuel, as follows:

Fuel consumption	1,800 lb.	2,000 lb.	2,200 lb.	2,400 lb.	2,600 lb.
Cubic feet air per pound pig..	44.2	49.1	54.0	58.9	63.8

Moisture in Blast.—The decomposition of the moisture in the furnace hearth absorbs heat according to the quantity present. Except for occasional leaks in water coolers, the only source of moisture in the hearth is that contained in the blast of air. The quantity of moisture in the air at any moment depends upon its temperature and the degree of humidity. According to the records of the United States Weather Bureau, the average temperature and humidity at Philadelphia, Pa., from 1874 to 1904 was as follows:

	Temperature. Degrees F	Humidity. Per cent.
Spring months	51	67
Summer months	74	70
Fall months	57	73
Winter months	34	74

Assuming an average for the State of Pennsylvania of 57 degrees F. temperature and 70 per cent. humidity, we have, from Davis's Meteorology, p. 143, that there are approximately 5.3 grains of water in each cubic foot of saturated air at 57 degrees F., of which 70 per cent. amounts to about 3.7 grains. The quantity of water in the air that is needed to consume the fuel which is required to smelt 100 pounds of pig iron may be found as follows:

$$\frac{\text{no. cu. ft.} \times \text{grs. per cu. ft.}}{\text{grs. per lb. av.}} = \frac{5300 \times 3.7}{7000} = 2.81 \text{ lbs. av.}$$

By the reaction, $H_2O + C = CO + H_2$, it appears that for 18 pounds of water in the form of steam which is decomposed, 12 pounds of carbon are oxidized to the condition of CO. One pound, H_2, burned to the condition of steam, gives 51,750 B. T. U. For each pound of steam, the heat given off will be $\frac{51,750}{9}$ = 5750 B. T. U., which will be absorbed on decomposition. The heat given off by the corresponding quantity of carbon, burned

to CO is $4450 \times \frac{12}{18} = 2965$ B. T. U. $5750 - 2965 = 2785$ B.
T. U , which is the net amount of heat absorbed per pound steam ;
$2785 \times 2.81 = 7826$ B. T. U., which is absorbed in decomposing
the moisture in the blast for each 100 pounds of pig made. This
amount of heat is equivalent to $\frac{7826}{4450} = 1.76$ pounds carbon
burned at the tuyeres, which equals 2¼ pounds coke of 78 per
cent efficiency, or 50 pounds coke per ton of pig.

The saving of such a small amount of fuel hardly appears to
be a sufficient incentive to efforts to desiccate the blast. Nevertheless this was successfully accomplished with great profit by
Gayley, who cooled his blast by means of refrigeration from 71 <small>Tr A
XXX'</small>
degrees F. to 25 degrees F., thereby reducing the moisture from
5.66 to 1.75 grains per cubic foot. The removal of 3.91 grains
per cubic foot is equivalent to about 63 pounds per ton of iron.
$\frac{63 \times 2785}{4450} = 39.4$ pounds carbon, or 50.5 pounds coke per ton
of pig. Yet the observed economy was far greater than expected.
The fuel consumption per ton of pig dropped from 2147 to 1726
pounds, and the daily output rose from 358 to 447 tons of pig.
The cumulative benefits of the reduction of fuel were very manifest The blast requirements fell from 40,000 to 34,000 cubic
feet per minute, with a consequent rise in temperature of 720
to 870 degrees F., the temperature of the escaping gases fell
from 538 to 376 degrees F , and they were of considerably less
volume, thereby carrying away less heat. The carbon ratio also
fell from 1.71 to 1.24. The summary below shows the change
very clearly.

	11 days natural blast, August 1-11 inclusive.	16 days dry blast, August 25-September 9 inclusive.
Average moisture	5 66 grs. per cu. ft.	1.75 grs. per cu ft.
Average daily output	358 tons	447 tons
Average fuel consumption	2,147 pounds	1,726 pounds
Average ore per unit of fuel	1.96	2.35
Number revolutions engine	114 per minute.	96 per minute.
Cubic feet air	40,000 per minute.	34,000 per minute
Temperature of blast	720 degrees F.	870 degrees F
CO in gases	22 3 per cent	19 9 per cent.
CO_2 in gases	13 0 per cent	16 0 per cent.
Temperature of gases	538 degrees F.	376 degrees F
Flue dust	5 0 per cent	1 0 per cent.
Indicated horse-power.	2,700	2,013
Indicated horse-power per ton coke	7 87	5.84

The credit for the most satisfactory explanation of this remarkable result belongs to J. E. Johnson, Jr., who points out that for every operation of smelting a certain critical temperature is necessary to perform the essential operations of the process, such as the reduction of the iron and metalloids, the fusion and superheating of the metal, the formation, fusion and super-heating of the slag, etc. This critical temperature corresponds closely to the melting point of the slag and varies with its basicity. The theoretical minimum quantity of heat needed by the furnace is the heat that will raise the products of combustion and contents of the hearth to the critical temperature, and it is only the excess above this point that is available for meeting contingencies, such as decomposition of water, etc. Any reduction in the extra heat requirements, as in decreasing the quantity of moisture, leaves an excess of heat above that required by the critical temperature, which permits of a corresponding decrease of fuel.

Tr. A. I. M. E., XXXVI, p. 1129

It has been pointed out by many, that the great saving resulting from the drying of the blast is due to the increased regularity of working, in consequence of which it is not necessary to carry an excess quantity of fuel when the atmosphere is dry, in order to have a sufficiency to meet the emergencies caused by varying degrees of humidity. As calculated by Richards, the saving of the 20 per cent. of fuel per 100 pounds of pig, is distributed as follows:

Tr. A. I. M. E., XXXVII, p. 355.

	Per cent
Less moisture to decompose......................................	2 95
Less wasted in gases...	5 25
Less radiation...	3 70
Less carbon ratio...	6 90
Less heat in slag...	0 10
Less heat in blast..	0.40
Total...	19 30

If the moisture could be held constant at any point, and the above equivalent of heat introduced in any form, such as increased blast temperature, it is evident that the same fuel saving could be effected with its attendant advantages.

The air for the blast is usually drawn from the engine room, and is therefore nearly saturated with moisture. The outside air rarely averages over 70 per cent. of saturation, and frequently gets down to 50 per cent. It is therefore desirable to draw the supply directly from out of doors. Even very moderate cooling of air has a considerable effect upon the quantity of moisture present.

> 1 cubic foot saturated air at 90 degrees F. contains 15 grs. vapor.
> 1 cubic foot saturated air at 70 degrees F. contains 8 grs. vapor.
> 1 cubic foot saturated air at 50 degrees F. contains 4 grs. vapor.
> 1 cubic foot saturated air at 30 degrees F. contains 2 grs. vapor.

For each drop of 20 degrees F., the saturation point is halved. Moreover, since the amount of vapor that can exist in a given

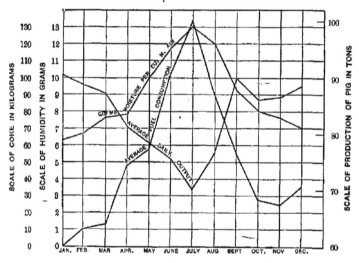

Diagram Illustrating the Relation Between Moisture in the Air Upon Fuel Consumption and Output.

space is dependent upon the temperature and independent of the presence of other gases, it follows that when at 15 pounds pressure 2 cubic feet of air are compressed into 1 cubic foot of space without change of temperature, there can be only one-half as much moisture to each pound of air. From these considerations, it is evident that even very moderate but uniform temperatures of air may work decided benefit in promoting regularity of working, through simple water cooling, and without expensive apparatus.

The accompanying diagram illustrates the effect of increased

Iron Age,
Mch 22, 1900,
p. 1032.

moisture in the air upon the fuel consumption and output of a blast furnace.

Volume of Gases.—When the blast comes into contact with the coke in the hearth of the furnace it burns to the condition of CO mixed with the residual nitrogen and hydrogen, whose volume, under atmospheric pressure at 60 degrees F. may be calculated as follows:

		Cubic feet.
CO, $169.56 \times 13.5 =$		2,289
N, $310.53 \times 13.5 =$		4,192
H, $0.31 \times 189.7 =$		59
Total		6,540

At the temperature of 3500 degrees F. the volume of these gases, if unconfined would be increased, as,

$$\frac{3500 + 458.5}{60 + 458.5} = 7.6 \text{ times, or}$$

$\frac{6540}{5300} \times 7.6 = 9.4$ times the volume of the original blast. Under furnace pressure, however, this volume is probably diminished by nearly one-half.

The volume of the waste gases as compared with the blast may be found as follows:

		Cu. ft.
CO_2	$106.40 \times 8.6 =$	915
CO	$136.16 \times 13.5 =$	1,838
N_2	$310.55 \times 13.5 =$	4,192
H_2	$0.31 \times 189.7 =$	58
H_2O	$17.75 \times 26.3 =$	467
Totals	571.17	7,470

which shows that the volume at the same temperature and pressure, as well as the weight, has increased to a little more than 1.4 times that of the blast. At the temperature of escape, however, the volume is in the ratio of their absolute temperatures, or $\frac{908.5}{518.5} \times 1.41 = 2.5$ times the volume of the original blast at atmospheric temperature and pressure.

Latent Heat of Expansion of Gases.—The expansion of the gases due to the sudden rise of temperature from 1200 degrees F. to 3500 degrees F. renders latent a considerable quantity of the

heat developed in the hearth. If the gases were allowed to expand
freely under atmospheric pressure in proportion to the increase
of temperature, we would expect them to render latent a quantity
of heat as follows:

B T. U
CO, 169.56 × 2,300 × 0 0715 = ...27,884
N₂, 310 55 × 2,300 × 0 0715 = ...51,070
H₂, 0 31 × 2,300 × 0 995 = ... 711

Total 79,665

which is equivalent to about 20 pounds of coke burned at the
tuyeres per 100 pounds of pig. However, as the pressure existing
in the hearth prevents free expansion, the heat rendered latent
by expansion up to the bosh probably does not greatly exceed half
that amount. A large proportion of the latent heat is returned
to the materials through the cooling and consequent contraction
of the gases as they pass upward through the furnace. The latent
heat of expansion above 60 degrees which is held by the gases as
they escape from the furnace at 450 degrees F. may be determined
as follows:

B T. U.
CO₂, 106 40 × 390 × 0 0455 = ... 1,888
CO, 136 16 × 390 × 0 0715 = ... 3,797
N₂, 310 55 × 390 × 0 0715 = ... 8,660
H₂, 0 31 × 390 × 0 995 = ... 120

Total...14,465

which is equivalent to about 4 pounds of coke burned at the
tuyeres, per 100 pounds of pig. It is evident, therefore, that the
latent heat rendered sensible again through the contraction of the
gases during their passage through the furnace, returns to the
stock all except this quantity of the heat originally rendered latent
through expansion of gases in the hearth.

Latent Heat of Expansion of Blast.—The heat rendered
latent by the expansion of the blast through its rise in temperature from 60 to 1200 degrees F. during its passage through the
stoves would amount to

$$407.75 + 1140 \times 0.069 = 32,070 \text{ B. T. U.}$$

if the blast expanded freely under atmospheric pressure. Since
the blast is under upwards of two atmospheres pressure, the
amount of expansion is about halved and the latent heat of expan-

sion of the incoming blast is, in consequence, probably about 16,000 B. T. U. We have seen that the latent heat of expansion lost in the escaping gases is something under 15,000 B. T. U. It is probable, therefore, that there is no net loss of heat to the furnace through heat rendered latent by expansion of gases, but, on the contrary, a slight gain. This gain, however, is more than neutralized by the heat rendered latent in vaporizing and expanding the moisture of the charge, thus:

B. T U.
absorbed.

Water 60° — Steam = 17.75 × 1,118 B. T. U =..................... . .19,845
Steam 212° — 450° = 17.75 (238° × 0.42 + 238° × 0.000185) =........ 1,775

Total........ ... .21,620

This absorption of heat is very efficient in keeping down the temperature of the furnace top.

REDUCTION OF METALLOIDS.

We have seen that the iron is easily reduced by gaseous carbon in the form of CO, and that its reduction is largely accomplished— very early in the process. The metalloids which combine with the iron to form the pig are not so easily affected, however, and consequently do not relinquish their original combinations until they are subjected to very strong persuasion at a considerable depth in the furnace.

Manganese.—The most readily reduced of the metalloids is probably manganese. It is usually present in the ore as dioxide, MnO_2, and it is probable that it is never completely reduced in the blast furnace. Under ordinary conditions, about two-thirds of the manganese in a pig iron mixture is reduced to the metallic condition and enters the iron, while the remaining third enters the slag and combines with SiO_2 in the form of MnO. Since its entrance into the slag is opposed by a slag of increased basicity, highly calcareous slags tend to drive the manganese into the iron. This disposition is favored, also, by high temperatures and strongly reducing conditions.

Silicon.—The silicon which enters pig iron is derived from the silica which is invariably present as gangue of the ores, impurity in the fluxes, and ash of the fuels. The quantity of silicon which enters pig iron usually ranges from 1 to 3 pounds per 100 pounds pig, while the quantity entering the slag may range from 5 to 15 pounds, but is usually about 7 pounds. Silica is not at all affected by CO and probably is not decomposed by solid carbon or other reagent even at the highest furnace temperature, except in the presence of a metallic matrix with which the silicon can unite at once, forming a silicide. The element is highly oxidizable and needs to form a protecting union immediately in order to preserve its elemental condition. The reaction evolves CO, and may be represented thus:

$$SiO_2 + 2C = Si + 2CO.$$

At temperatures below 2700 degrees F., silicon will absorb oxygen at the expense of the oxides of carbon, and it is only at temperatures well above this that the preferential relation is reversed and reduction of silicon by carbon becomes possible. It is evident then, that reduction of silicon can take place only within the melting zone. We have seen already that the silica of the gangue and flux are melted at temperatures about 3000 degrees F., and having once entered the slag, the silicon is probably secure from reduction. It is highly probable, therefore, that the bulk of the silicon is derived from the only remaining source, namely the ash of the fuel. This probability is well substantiated by the fact that siliceous fuels always produce higher silicon iron than when the silica of the charge exists mainly in the gangue of the ore.

The reaction, $SiO_2 + 2C = Si + 2CO$, represents a considerable absorption of heat, thus

	B T U
Heat absorbed in reducing 1 pound Si =	11,571 5
Heat given out by corresponding amount of carbon, $\frac{24}{18} \times 4,450 =$	3,814 5
Net heat absorbed per pound Si =	7,757 0

$7757 \times 1 5 = 11,635$ B. T. U. absorbed by the reaction, per 100 pounds pig.

Phosphorus.—Phosphorus enters the furnace in all members of the charge. It is usually in the form of calcic phosphate, called apatite, $Ca_3P_2O_8$. This compound is not affected by CO, and is probably not dissociated by heat alone, in any part of the furnace. In the presence of silica, at slag-forming temperatures, however, it can release its CaO to the slag, leaving the P_2O_5 free for reduction. As the reduction of phosphorus from its oxide is difficult, it is probably accomplished by solid carbon, thus:

$$P_2O_5 + 5C = P_2 + 5CO.$$

For ordinary charges and furnace conditions, this reaction is complete, and practically all of the phosphorus in the charge enters the iron. Exceptional conditions, such as a very highly phosphoric charge, or a highly basic or ferruginous slag, may result in some elimination of phosphorus, but it is safe to assume that ordinarily all of the phosphorus charged in ore, fuel and flux will enter the iron. The reduction of phosphorus is an endothermic reaction also, as shown thus:

	B. T. U.
Heat absorbed in reduction of 1 pound P =	10,605.6
Heat generated by equivalent amount of carbon, $\frac{60}{62} \times 4,450 =$	4,306.4
Net heat absorbed per pound P =	6,299.2

$6299.2 \times 0.5 = 3149.6$ B. T. U., absorbed by reaction per 100 pounds pig.

Sulphur.—Sulphur enters the furnace in a variety of ways. It may be present in the form of either sulphate or sulphide in each member of the charge, but usually at least 90 per cent. of that present is found in the fuel. It exists in coke in three different conditions: a small portion as sulphide can be evolved as H_2S, a small portion can be separated as sulphate of metals, and the remainder, ranging from 65 to 90 per cent., exists in combination with the carbon. Ordinarily the quantity of sulphur in the charge does not vary much from 1 pound for every 100 pounds of pig made, and of this quantity 2 to 10 per cent usually finds its way into the iron. The bulk of the remainder enters the slag, while a small percentage is volatilized.

Since most of the sulphur is held by the carbon, it follows that it is not released until the carbon is burned before the tuyeres. The greater part is burned to SO_2, but about 15 per cent. escapes as SO_3. Its volatility was found by Wuest and Wolff to differ with the temperature, and the nature of the atmosphere. The evolution is most nearly complete in the presence of hydrogen or steam, less complete with CO and CO_2, and least satisfactory with nitrogen. The action in any atmosphere is accelerated by higher temperatures.

Inst. Jour.,
1905, I.,
406

	Per cent. sulphur present volatilized in the presence of				
	H₂.	Steam.	N₂.	CO.	CO₂.
930 degrees F	7.50	12.84	2.41	12 80	6 47
1,110 degrees F	22 99	13 27	4.90	16.89	8 32
1,470 degrees F	41.87	36.83	5.90	30.80	16.00
1,650 degrees F	45.77	51.52	6.97	37.61	25 46
1,830 degrees F	51.17	54.34	17 35	38.32	59 24

By passing a mixture of the above gases of the following proportions:

N₂.	CO.	CO₂.	H₂
56.0	28	13 5	2 5

over Fe_2O_3, $CaCO_3$ and a mixture of the two, the following comparative results were obtained:

	Per cent. of total sulphur which was volatilized.			Per cent. of volatilized sulphur which was absorbed			
	Fe_2O_3	$CaCO_3$.	Fe_2O_3+ $CaCO_3$.	Fe_2O_3	$CaCO_3$.	Fe_2O_3+ Fe_2O_3	$CaCO_3$ $CaCO_3$.
480 degrees F	44.32	61.74	41.51	53.06	10 26	56.51	0
930 degrees F	49.31	59.94	33.31	64 02	16.68	59.44	0
1,110 degrees F	47.17	53.47	31 17	68.33	47.05	59.32	2.34
1,470 degrees F	58.67	42.66	40.49	69.58	68 08	18 28	45.78
1,650 degrees F	58 39	40 91	31.27	68.02	85.74	41.03	43.00
1,830 degrees F	53.64	40.57	34 47	62 66	91.83	0.0	100 64

At low temperatures, Fe_2O_3 absorbs sulphur and is partially reduced by it. At low temperatures, $CaCO_3$ is not active in absorbing sulphur. At 1100 degrees F., however, $CaCO_3$ becomes more active and at 1800 degrees F., the absorption is practically complete. This is because $CaCO_3$ begins to decompose at 1100 degrees F. The action is not strong, however, until the temperature approaches 1500 degrees F., when it begins to cohere At 1650 degrees the reaction becomes violent and at 1800 degrees it is practically complete. It appears, then, that sulphur does not reach the hearth in its original condition, but is volatilized and passes upward to be absorbed by members of the charge and brought back to the melting-zone again. At temperatures below 1100 degrees F., it is absorbed chiefly by Fe_2O_3, but at temperatures above 1650 degrees F., the bulk is absorbed by the $CaCO_3$.

Since sulphur is an acid radical, whether it is in the oxidized or elemental condition, it combines readily with bases of the slag, such as lime, and the more basic the slag, the more strongly it will be held. Under ordinary conditions, it unites with lime to form a sulphide of calcium, which is dissolved in the slag.

It is the universal experience that sulphur is best excluded from pig iron in the presence of a basic cinder at high temperatures. This fact may be accounted for in several ways. It was formerly considered that the high temperature volatilized the sulphur, but owing to condensation in cooler parts near the top of the furnace, more or less of the sulphur is returned to the hearth, therefore this theory does not hold. It has been shown recently that sulphur forms with silicon a sub-sulphide that is volatile, which may account for its elimination from hot furnaces where much silicon is reduced. High silicon irons are generally noticeably free from sulphur, and this fact has been attributed to their incompatibility. But it is more likely that the reduction of silicon leaves the slag more basic and hence better able to retain sulphur. In the presence of metallic manganese, sulphur unites to form a stable sulphide which enters the slag, leaving the iron comparatively free from sulphur.

Jour. Am.
F d'v'n Asssoc.,
1903, p. 317.

CYANIDES.

The action of the nitrogen of the blast upon the hot carbon in the hearth results in the formation of small quantities of cyanogen, CN. Cyanogen shows a decided affinity for the sodium and potassium which are usually present in small quantities in the fuel ash, and as a result, cyanides of the alkali metals are found in the gases in all parts of the furnace. Bell found that the average quantity in a coke furnace for six consecutive days was 6.58 grains of cyanogen per cubic feet of gases in the hearth, and 1.65 at the top, or about 130 and 33 pounds respectively per ton pig, which would indicate at least double these quantities of the mixed sodic and potassic cyanides. The action of the cyanides upon the members of the charge is not well understood. It is known, however, that they are strong reducing agents, since they take up oxygen and form cyanates. Experiments by Bell show that they are more active than CO in the presence of CO_2, thus:

Inst Jour,
1883, II,
p 555.

Gas mixture.		Temperature, degrees F.	Time, hours.	Per 100 parts iron present.		
Vol. CN.	Vol. CO_2			Fe as metal	Fe as oxide.	Carbon deposited.
1	6	1,288	2¾	56.3	43.7	28 50
1	15	1,492	2½	6 5	93.4.	1.30
1	30	1,427	3	0.9	99.1	2.52

Ibid,
p. 556.

This shows that cyanides are very active, even when CO_2 is six times their volume, which would quite neutralize CO.

For this reason some metallurgists attribute great importance to the cyanides as reducing agents. Their action may be represented by the following equation,

$$KCN + FeO = Fe + KCNO.$$

The resulting potassic cyanate passes upward in the current of gases and is probably decomposed by CO_2 into carbonates with the liberation of the nitrogen, thus:

$$2KCNO + CO_2 = K_2CO_3 + CO + N_2, \text{ or}$$
$$KCN + KCNO + CO_2 = K_2CO_3 + 2CN.$$

The potassic carbonate is undoubtedly deposited in the cooler parts of the furnace and carried downward by the descending materials to the hearth, where it can again act as a base for cyanogen In this way, the quantity of cyanides, which depend for existence upon the presence of alkalis, is cumulative, and is said to amount sometimes to as much as 2 cwts. of cyanogen per ton of iron. The removal of the last traces of oxygen from the ore in the descending charge is credited by some to the action of cyanides instead of to the deposited carbon. The fact that reduced, spongy iron shows little or no carbon when it reaches the melting zone, lends color to this theory. However, 130 pounds per ton is equivalent to 5.8 pounds CN per 100 pounds pig, and is capable of removing 3.55 pounds of oxygen from the ore, which is only about 80 per cent of that usually removed by solid reagents. It is probable, therefore, that the reduction is completed by the joint action of carbon and cyanides The gradual accumulation of cyanides in a blast furnace may partially account for the continuous improvement in its action during the first few months of a blast. Roberts-Austen, Metallurgy, p. 195

A considerable quantity of cyanides escape from the furnace in fume and flue-dust, and owing to their extremely poisonous character should be handled carefully. In 1901, 20 tons of flue-dust, deposited in the River Ems in Styria, caused the death of a great number of fishes, for which the furnace company was fined $14,000. In 1898, a similar accident happened in the Rockaway River at Wharton, New Jersey. Inst Jour, 1904, II, p 540

ANOMALIES IN GAS COMPOSITION.

The refixing in the form of cyanides and carbonates of the carbon, which has been gasified at the tuyères causes a temporary withdrawal of both C and O from the gases in the middle of the furnace. Numerous analyses of the gases by Bell gave this general result:

Place of sample.	C.	O.	O expected
Hearth	20.43	27.74	27.24
Just above bosh	.18 25	26 01	26.28
Top	20.96	33.40	36 15

(*Inst Jour, 1887, II, p 82*)

It may be observed that the quantity of oxygen in the hearth is slightly in excess of the nitrogen as well as of the carbon. This is due to the fact that some of the oxygen comes from the moisture in the blast and from the reduction of metalloids while some of the nitrogen is withdrawn to form cyanides. By the time the gases have reached the bosh, they show a distinct deficit of both carbon and oxygen, that of the latter being somewhat more pronounced. This is no doubt due to the fact that the atomic equivalent of oxygen in cyanates and carbonates is greater than that of carbon. It is unsafe to speculate, however, on the conditions of equilibrium, since, as we have already seen, both carbon and oxygen are deposited from the gases through the influence of the reduced iron sponge. That the carbon, oxygen, and nitrogen are more or less perfectly restored to the gas current, is accounted for by the decomposition of the alkaline cyanates and carbonates in the upper part of the furnace. The imperfections of the restoration may be at least partly explained by the fact that solid matter is carried out of the furnace in the " fume," which is borne away by the current of gases.

Fume.—The fume is the finely divided particles of solid matter which float in the gaseous current, giving it its characteristic whitish appearance as it issues from the furnace and chimneys. The appearance of the fume varies with the quantity of matter carried, being white and dense when the furnace is hot, and thin and bluish when the furnace is cold. It is therefore an indication of the internal condition of the furnace. Some of the material which exists in the gases in the hotter parts of the furnace is condensed before the gases escape and hence does not appear in the fume, while other material is caught up early in the descent by the current and so never reaches the lower part of the furnace. The former diminishes as the gases pass upward, while the latter increases. The following analyses, selected from those of Bell, at Clarence, England, illustrate the changeableness of composition of the solid constituents of the gases while in the furnace:

Distance above tuyeres	2 ft	26½ ft.	30¾ ft	45½ ft.	58¼ ft.	76 ft.	
Probable temperature	3,500° F	2,000° F.	1,800° F.	1,600° F.	900° F.	600° F.	
Grs fume per cubic foot gases	5 95	3 38	0 09	0 61	0 72	0.83	
KCN	75 98	89 20	71 21	...	...	...	
NaCN	...	3 51	0.07	...	...	...	
K₂CO₃	3 05	. .	. .	. .	...	} 2.21	
Na₂CO₃	20 70	3 52	3.91	. .	...		
NH₄Cl	...	...	...	...	..	2.39	
KCl	0 55	. .	.	5 42	1 80	5.92	
NaCl	...	2 48	6 20	2.93	1 00	1.48	
ZnO	Tr	Tr.	6 71	15 27	17 17	13 39	
PbCl₂	. .	. .	...	1 47	8 77	...	Bell's "Principles" of
CaSO₄	. .	. .	...	Tr.	0 57	3 60	Manufacture of Iron and Steel,
SiO₂	.	...	...	6 38	15 00	10 25	p 225.
CaO	...	. .	...	6 70	14 37	0 56	
Al₂O₃	...	. .	...	1 84	6 11	3 66	
MgO	.	. .	...	0.31	4 33	Tr	
FeO	.	...	...	1.03	2 63	...	
Fe₂O₃	. .	...	...	. .	. .	15 00	
C	..	...	...	0 27	0.41	13 66	
Composition of gases.							
CO₂	1.90	0.46	0.00	3.84	6 56	10 69	
CO	39 18	35 80	34 82	32 33	32.25	28 58	
H₂	1.97	1 14	1 41	3 88	1.08	1.70	
N₂	56 95	62 60	63 77	59.95	60.11	59.03	

REDUCTION BY CHARCOAL.

The process of reduction of iron ore in a charcoal furnace appears to be of a nature somewhat different from that in a coke furnace. Whereas, in coke furnaces reduction begins almost as soon as the ore is charged and is practically complete at a depth of 20 feet, in charcoal furnaces, which are usually of much less height, the reduction has hardly begun at that depth A comparison of the composition of the gases in an 80 foot coke furnace and a 37 foot charcoal furnace shows the difference very clearly

	Coke furnace.		Charcoal furnace	
	CO	CO₂	CO	CO₂
	CO	CO$_2$	CO	CO$_2$
Escaping gases....	29 5	11.0	17 18	15 42
16½ feet from top..	34.1	2 2	...	...
18 feet from top ...	...	...	19 23	14 76
39 feet from top........	35.0	1 1	..	...
25½ feet from top. . .	..	...	18 44	14 79
65 feet from top	35 9	0 5	. .	...
32 feet from top ,top....	.	...	20 92	13 29
70½ feet from top	36 6	0 0	...	...
34½ feet from top...	...	...	26 33	8 20

The small percentage of CO$_2$ at depths below 16 feet in the coke furnace shows that little reduction takes place in those levels. The percentage of CO$_2$ in the gases of the charcoal furnace shows clearly that reduction is not very active until a depth of 32 feet is reached Between that level and the hearth, the reduction is very active and is necessarily very sudden, as over 80 per cent. takes place in the last five feet of the journey. In some instances it has been observed that reduction in the charcoal furnace had not even begun at a depth of 30 feet. Not only is the action postponed, but its actual occurrence is different. We have seen that the action of CO in the coke furnace produces metallic sponge direct. In the charcoal furnace, metallic iron does not appear until the reduction has proceeded for some time and the lower oxides of iron form intermediate steps in the operation as shown by experiments by Ebelmen, who lowered ores into the furnace in a closed box which was withdrawn for examination · ·

Time,	Depth,		First experiment.			Second experiment		
hours.	feet	Temperature.	Fe₂O₃	FeO.	Fe.	Fe₂O₃.	FeO.	Fe.
hours.	feet	Temperature.	Fe_2O_3	FeO	Fe.	Fe_2O_3	FeO	Fe.
2	8	Black . .. .63 4		3 2	0 0	37 0	Tr	0 0
4½	14½	Dull red 33 0		32 5	0.0	27 8	12 7	0 0
5½	16½	Cherry red 26 0		41.8	Tr	24 1	17 3	0 0
6½	18¾	W I softens ... 0 0		35 0	26 7	0 0	30 2	10.0

Fuel Consumption.—The fuel consumption in a charcoal furnace is usually less than that in a coke furnace, even when· cold blast is used. This is largely due to the fact that since the ratio of $\dfrac{CO}{CO_2}$ is usually very low, the fuel is better utilized. Such a low ratio is made possible by the fact that the ore does not depend entirely upon CO for its reduction. It is to be noticed that reduction takes place at temperatures where solid carbon can take up oxygen from the ore. Morover, the quantity of CO_2 in the top of the furnace is sufficient to retard reduction greatly, while it is usually too small to be formed by the action of CO alone on the ore. These facts point to the probability that there is a considerable amount of reduction of ore by solid carbon in the lower furnace and that the resulting CO performs additional reduction in the cooler parts of the furnace, thereby doing double duty. An additional cause for the low fuel consumption rests in the fact that the lessened volume of blast needed, owing to the reaction between the solid carbon and fixed oxygen, permits slower travel through the cold stock and hence more thoroughly cooled gases. Owing to the lack of reduction in the top of the furnace there is no evolution of heat there. The temperature of gases escaping from a charcoal furnace has been recorded as low as 117 degrees F. for a week at a time. Finally, the rapid driving of charcoal furnaces permits the passage of stock through the furnace at such a rate that loss by radiation per ton of product is materially lessened.

SOURCES OF ECONOMY.

The function of the blast furnace is the recovery of the iron in the charge with as little expenditure as possible. The smelting charges are made up of the cost of the materials, and the cost of handling them and the products. The cost of materials and labor are fixed by the natural conditions and do not allow much latitude to the furnacemen. The quantity of materials used, except, perhaps, in the case of fuel, bears a constant ratio to the output. There is, therefore, not much opportunity for saving, except in the consumption of fuel.

The quantity of fuel used in smelting iron always depends upon several factors. It is influenced, primarily, by the degree of oxidation to which its carbon is burned. The smaller the ratio $\dfrac{CO}{CO_2}$, the greater the quantity of carbon that is burned to CO_2, by which the maximum quantity of heat is developed. In the second place, the more effectively the products of combustion are

cooled, the more of the heat developed will serve a useful purpose
Well-cooled gases can be attained only by having a long pathway
of cold materials through which they move slowly. Thirdly
since the losses of heat by radiation and conduction are function
of time and not of yield, the more rapidly the materials pass
through the furnace, the less will be the loss of heat per ton
This condition can be best attained by rapid driving. Finally, the
more heat that can be introduced into the furnace, without the
aid of fuel, the less fuel will be needed. Hence, a well heated
blast is conducive to fuel economy.

The effect of a change in any one of these factors is cumulative. It disturbs the equilibrium that exists between them and
compels readjustment. Any cause that leads to a decrease in fuel
used, necessarily lessens the quantity of blast needed, and hence,
also, the quantity of escaping gases. In consequence, the gases
move more slowly, give up more heat to incoming materials, and
pass off at a lower temperature. Since there is less carbon in the
gases, there will be less escaping CO. As the quantity of CO_2 per
ton is fairly constant, less CO means a lower ratio, $\dfrac{CO}{CO_2}$. The less
the quantity of blast, the more thoroughly it will be heated by the
heating arrangements, which brings a further reduction in fuel.
The less blast needed per ton of pig, the more tons can be made by
a given capacity of equipment in a given time and the less will be
the radiation and conduction losses. For these reasons, a seem-
ingly slight change of conditions frequently brings a surprisingly
large difference in results.

Effect of Furnace Height.—Other things being equal, the
height of a furnace has a profound effect upon its economical
operation. The first result of an increase of height of a small
furnace is a more thorough cooling of the escaping gases. This
will permit a lessened quantity of fuel and flux, hence, less slag
to be melted, and more thorough combustion of the carbon. The
increase of blast temperature gives identical results. The effects
of increased height of furnace and increased blast temperature
are shown by the following comparative table:

	(1)	(2)	(3)	(4)	(5)
Height of furnace ...	48 feet.	48 feet.	80 feet.	80 feet.	90 feet.
Cubic feet capacity .	6,000	6,000	15,000	15,000	33,500
Temperature of blast..	770° F	905° F.	905° F	1,300° F.	1,310° F.
Temperature of gases..	844° F.	806 °F.	590° F	482° F.	432° F
Wt. coke per ton pig	40 76 cwt.	28 50 cwt.	22 50 cwt	19 99 cwt	19 69 cwt
Wt. stone per ton pig	18 25 cwt.	16 00 cwt.	11 00 cwt	11 00 cwt	10 50 cwt.
Wt ore per ton pig...	46 20 cwt	48.00 cwt.	48.00 cwt.	48 00 cwt.	48.00 cwt.
Wt slag per ton pig..	34 15 cwt	31.00 cwt.	28.00 cwt.	28.00 cwt.	28.00 cwt.
Wt blast per ton pig.	180 46 cwt	126 05 cwt	100 41 cwt.	87 15 cwt	84 01 cwt.
Wt. gases per ton pig	233 92 cwt.	168 37 cwt.	135 20 cwt.	119 47 cwt.	115 89 cwt.
Ratio gases to blast .	1 296	1.336	1 346	1 371	1.378
Output	90 tons.	220 tons.	350 tons	350 tons	700 tons.
Heat requirement....	111,180 cal.	104,336 cal	91,194 cal	88,577 cal.	85,912 cal.

By comparing (1) with (2), and (3) with (4), it may be observed that increased blast temperature in otherwise similar furnaces resulted in decreased fuel consumption and consequently a decrease in blast, flux, waste gases, and slag. By comparing (2) with (3) and (4) with (5) it may be seen that increase in height of furnaces results in increased output, cooler gases and lessened fuel, flux, blast and waste gases Of course, added height will be of no avail unless the furnace be kept full. Allowing the stock to settle will result in increased fuel consumption.

Limits of Size.—It would seem, therefore, as if a continual increase of height, ad infinitum, would result in a continued fuel economy Such might be the case if it were not for the fact that the complete cooling of the gases is rendered impossible because the reduction of the ore is exothermic in nature. It is consequently useless to attempt to cool the gases below the temperature that will be attained by the process-of reduction. It is evident, therefore, that the maximum saving of waste heat will be reached when the column of intercepting materials between the hearth and the point where reduction by CO ceases, is long enough to cool the gases to the temperature of the reaction; in other words, when the zone of interception is sufficiently long to remove the zone of reduction beyond the influence of the hearth temperature. From the above comparisons, Bell concluded that there was no advantage to be gained by increasing the height of furnaces beyond 80 or 90 feet, and recent experience with furnaces 100 feet in height has amply confirmed his conclusions.

It is only logical to assume that the size of a furnace should bear a definite relation to the hearth area, or at any rate to the amount of work done per square foot of hearth area. The object of size in a furnace is to enable a large quantity of stock to be presented to the hot gases in order that as much of their heat as possible may be extracted It is evident, then, that height alone, or breadth alone, does not determine the proper relationship, since it is manifestly a question of volume A moment's consideration, moreover, will reveal the fact that of all the heat used in the furnace, that portion only which is absorbed by the charge can apply toward the melting of the iron and slag. Hence, if we can cause the charge to intercept enough heat during its passage through the furnace to fuse the products, that is all we can hope to accomplish. The proper furnace volume per square foot of hearth area will evidently vary with the quantity of heat developed, but for a given degree of activity the volume may be determined approximately by this formula: The number cubic feet of stock needed per square foot of hearth area =

$$\frac{\text{Heat capy. of products per sq. ft. of hearth area in 24 h.}}{\text{Heat capy. of the charge per cu. ft.}}$$

Let us assume, for example, that a furnace burns 6000 pounds of coke per square foot of hearth in 24 hours, thereby producing 6000 pounds of pig and 55 per cent. as much slag; then the amount of heat which can reside in the molten products of one square foot of hearth area in 24 hours will be:

	B. T. U.
$6,000 \times 0.22 \times 2,700 =$	3,564,000
$3,300 \times 0.30 \times 3,000 =$	2,970,000
Total	6,534,000

which is the total heat capacity of the molten product of one square foot of hearth area per 24 hours and is the numerator of the fraction.

The heat capacity of the charge per cubic foot is evidently the heat capacity per pound multiplied by the weight per cubic foot. The weight of the charge per cubic foot may be found as follows:

100 pounds coke occupies	3 57 cu. ft. space.
170 pounds ore occupies	1 13 cu. ft. space.
50 pounds stone occupies	0.50 cu. ft. space.
320 pounds charge occupies	5 20 cu. ft. space.

Each cubic foot weighs $\dfrac{320}{5.2}$ or 61.5 pounds when it enters the furnace. In the early part of the descent, however, the ore loses its water and much of its oxygen; the coke loses some carbon through solution by CO_2; the stone loses some CO_2 through decomposition. During the greater part of the descent, therefore, the weights of the members of the charge will be modified as follows:

	Pounds.
100 pounds coke, less 5.7 per cent. carbon stolen, equals	94 3
170 pounds ore, less 30 per cent O_2 and H_2O, equals	119 0
50 pounds stone, less 42 per cent CO_2 evolved, equals	29.0
320 pounds charge is modified until it equals	242 3

$61.5 \times \dfrac{242.3}{320} = 46.6$, which would be the weight per cubic foot of the descending materials if they occupied the original space. As they compact somewhat upon changing form the final weight per cubic foot is probably not far from 50 pounds.

The heat capacity of the charge per pound is as follows:

	Pounds			B. T. U.
Iron	1 00	× 0.2	× 2,700 =	540 0
Gangue	0 55	× 0 3	× 3,000 =	495 0
Coke	0.943	× 0 428	× 3,000 =	1,211.7
	2 493			2,246 7

Whence $\dfrac{2246.7}{2.493} = 900$ B. T. U. absorbed per pound of charge.

$900 \times 50 = 45,000$ B. T. U. absorbed per cubic foot, which is the denominator of the fraction. Whereupon $\dfrac{6,534,000}{45,000} = $ 145, which is the number of cubic feet of charge needed to absorb the quantity of heat that will suffice to fuse the products of 24 hours from one square foot of hearth area. If the furnace has a bosh area two and one-half times that of the hearth, the height of the column of stock which contains 145 cubic feet will be $\dfrac{145}{1.75} = 83$ feet.

For a furnace which passes its own volume of material every 15 hours, this would require a working column from melting zone to reducing zone of $\dfrac{15}{24} \times 83 = 52$ feet. Assuming that the zone of fusion is 15 feet above the hearth level and that the reducing zone is 8 feet deep, and the bell clearance 10 feet more, the total height of the furnace should be $15 + 52 + 8 + 10 = 85$ feet, which is found ample in practice. The volume of stock which should pass the hearth in 24 hours is greatly affected by the rate of combustion and the rapidity of movement of the stock. With the very low rate of combustion of 3000 pounds per square foot, as in the case of anthracite, the heat intercepting zone need not be over 20 feet high, and the furnace is therefore required to be only 50 to 55 feet in height. In the case of charcoal, where the furnace may empty itself several times in 24 hours, a lesser height is necessary than in a coke furnace to present the requisite quantity of stock to cool the gases satisfactorily.

Rapid Driving.—Other things being equal, the rapidity with which materials pass through the furnace is in proportion to the quantity of blast which enters the furnace. The prime requisite for rapid driving is that the fuel should be porous, so that it may unite readily with the blast and develop rapidly the heat for melting. In the second place, it is necessary that the charge should be permeable to gases, and the ores readily reducible, in order that they may come down to the hearth in the proper condition for melting. The furnace hearth should be large, so that the materials may readily present themselves for fusion, and the bosh should be steep to prevent any sticking or hesitancy in the descent of materials in the zone of fusion. Finally the ores should be rich in iron, in order that too much space may not be occupied by unproductive substances.

Regularity.—The state of affairs that probably conduces most to fuel waste is irregular working. It may be conveniently considered as synonymous with uneven distribution of heat from point of either time or place. If the proper quantity of fuel is charged in a furnace, and, for any cause, an undue proportion of it arrives in the zone of combustion at one time, there will evidently be at some subsequent time a deficit which must be corrected by extra additions Or, if through a difference of permeability, the gaseous current is directed to one part of the charge to the exclusion of the others, or if, through the breaking of water connections or other causes, local cooling is set up, the ultimate corrective must be the supplying of the deficit by additional fuel For these reasons, the materials which compose the charge should be introduced with due regard to uniformity of composition and regularity of descent.

Stock Distribution.—One of the chief causes of difficulty in maintaining uniformity of distribution and descent of stock is the mixing of coarse and fine materials. When a mixture of materials of differing sizes is poured in a heap, it usually happens that the finer portion forms a cone, whose slopes make an angle of about 40 degrees with the horizontal, while the lumps roll down and form a ring at the base. Applying this principle to the filling of a blast furnace, we see at once that if the distributing bell is too large, or if the stock gets low in the furnace, the charge will be deposited against the walls and the lumps will collect in the hollow at the center and form a central column of undue permeability. On the other hand, a bell that is too small will deposit the material in a ring, and the lumps will roll to the walls as well as toward the center, leaving an annular column which will be relatively impervious to the gases. The ideal condition would be to distribute the materials so that they would be uniformly mixed. The best effects are obtained when the distributing bell has a diameter 4 or 5 feet less than the stockline, thereby leaving 2 to 2½ feet clearance on all sides of the bell.

CHAPTER VI.

FURNACE IRREGULARITIES.

A blast furnace in operation is subject to many forms of irregularities, which are due to a variety of causes. As a rule,
such irregularities need to be detected promptly, diagnosed accurately and acted upon immediately, in order that the furnace output may not suffer, or, indeed, the safety of the furnace may not
be endangered. The ability to recognize the first symptoms of disorder and thereby to counteract or prevent the train of evils that
usually follows, is an invaluable asset to the successful furnace
manager. This ability is acquired only by long experience and
close attention, supplemented by a natural aptitude for discern-
ing that which is not always distinctly indicated.

LEAKY TUYERES.

The most common trouble to which a furnace is subject is
probably leaky tuyeres. A leaky tuyere results usually from
wear or from local superheating or "burning" of the metal,
whereby the cooling water is allowed to leak into the furnace
hearth. If the leak is not promptly discovered and stopped, it
may result in serious cooling of the hearth. The leak may occur
in any part of the tuyere, but the most vulnerable point is the
tip of the nose, and especially the upper surfaces. It may be due
to abrasion of the constantly descending stock during a long
period of service, which wears the metal so thin that it cracks
under the strain. More frequently, probably, the damage is
caused by a stoppage of the flow of cooling water, whereby the
enclosed water is converted into steam at the hottest point, and
the integrity of the metal destroyed through superheating. Another cause of burning is the alloying of molten iron which drops
on the tuyeres in its passage toward the hearth, or is directed
toward them too constantly by some obstruction. The burned
tuyeres that so frequently follow the dislodgment of scaffolds
probably result from the splashing against them of molten iron.

234

Detection of Leaks—There are several ways in which a leaky tuyere becomes manifest Occasionally it is announced by a loud report like an explosion, or by the increased volume and inflammability or " wildness " of the escaping gases. More often, however, it is suspected from its effect upon the escaping products of fusion. The cinder becomes dark, shows a black·crust or appears foamy or glassy. At the same time the sulphur in the iron runs up, showing that the ˙hearth is abnormally cool. When a leak is suspected, its existence may be verified by investigation. Sometimes dampness can be seen about the base of the tuyere or cooler. A cold bar thrust in at the peep hole will generally show dampness when withdrawn if there is a leak. Tongues of blue flame may break through the walls about the tuyere. Sometimes, water can be seen by looking into the tuyere, particularly if the blast be thrown off. If the water inlet of the tuyere be shut off, the gases from the hearth will work into the leak and can be ignited by a torch at the discharge pipe.

Changing Tuyeres.—When the presence of a leaky tuyere has been established, no time should be lost in replacing it. The usual method of extracting a tuyere is to stop the blast, remove the blowpipe, insert a tuyere hook, and by means of a claw in the hands of several lusty helpers, to jar the tuyere loose. Occasionally it becomes so expanded by heat while leaking that it binds in the cooler and vigorously resists extraction. Sometimes this condition may be overcome by slacking the water in the cooler and allowing it in turn to expand enough through heating to release the tuyere. As a last resort it may be necessary to melt the tuyere completely out by means of an oil blowpipe. As soon as it is removed, a new one which has previously been filled carefully with water and all air removed, is thrust into place, wedged tightly by blows from a dolly and the water connections promptly made. The blowpipe is then replaced and the blast turned on. Every minute spent in replacing a tuyere is dead loss to the furnace, since the blast must be stopped entirely during the change. The time of changing should not be over 6 minutes, but frequently takes 10 to 20.

Causes of Burning.—Ordinarily a well cooled tuyere with a positive water circulation will resist the heating effect of small quantitics of molten iron. Molten slag will not affect bronze unless it carries shots of iron. There are several ways in which the dripping iron may burn the tuyeres. Sometimes the molten material cuts grooves above them, or is deflected by some obstruction, so that a small stream of iron impinges constantly on the same spot in the tuyere, thereby weakening it and heating it to the steaming point. Lumps of fuel or chilled cinder lying before the tuyeres may deflect the iron against them and cause damage. A persistent obstruction should be removed by a pricker. Ores containing lead or zinc may cause trouble because of the readiness with which those metals alloy with bronze.

If the blast enters through the tuyere under good pressure and penetrates directly to the hearth center, there is little action immediately above the tuyeres and an accumulation of material frequently forms on the nose of each tuyere and protects it from the dripping iron. If the water supply is suitable, the repeated loss of a tuyere indicates irregular working.

The sudden stoppage of the blast when the hearth is full is always a source of danger to tuyeres, since it allows slag and metal to flow into them when the pressure is removed. Before throwing off the blast, the cinder should always be flushed if possible. If it is near casting time, it may be advisable to open the tapping hole also.

Prevention of Burning.—There appears to be no regular means of preventing burned tuyeres, except when they are caused by water stoppage. Water stoppage is due to obstructing particles carried in the water supply. A tolerably pure water should be used, and passed through a screen to remove sticks, leaves, fishes, etc., before it enters the tuyeres. A pressure of 25 pounds per square inch will generally suffice to keep fine sediment in motion. A higher pressure of water or steam should be used periodically to wash out any accumulations that may have formed. This should be done regularly every day or week according to the quality of water used. Higher pressures should not be used regularly, as experience shows that tuyeres that are successfully protected by a pressure of 25 pounds will burn frequently with 40 pounds. It is suggested that water at such high

Iron Age, Jan. 24, 1895.

Iron Age, Feb. 21, 1901.

pressure striking against the nose of the tuyere, bounds away without allowing sufficient period of contact to cool it properly

Other Coolers.—**Bosh-plates** and **tuyere coolers** are subject to similar vicissitudes, but their destruction is far less frequent, as they are less exposed to attack from the molten iron. The **cinder-notch cooler** and **intermediate cooler** are also comparatively safe from trouble. The **monkey,** however, is occasionally attacked by a molten iron which may present itself at that level with a viscous cinder or when, for any reason, casting has been delayed too long. The changing of the monkey, however, is a comparatively simple operation, and is subject to the same principles as the changing of the tuyeres, as is also the changing of any other water-cooled part of the furnace

DESTRUCTION OF LINING

The destruction of the lining at any point in the furnace above the melting zone will result in the overheating of the shell, and the formation of what is commonly called a " hot spot." If the lining is destroyed within the melting zone, the molten material may eat through the outer shell and a " break-out " result.

The lining of a blast furnace is subject to a great variety of conditions throughout its course. Its temperature ranges from a little above boiling water to the melting point of platinum, and its surface is subjected to the abrasion of the countless angles of solid stock and to the fluxing and disintegrating action of thousands of tons of molten matter, as well as the corrosive action of the gases. According to Lürmann, there are four distinct agencies at work tending to destroy the furnace lining

(1) The abrasion of the descending solid materials, which affects all depths above the zone of fusion.

(2) The action of certain constituents of the gases, notably CN and its salts, which are volatile at depths below 20 feet.   Inst Jour, 1892, I, p 397

(3) The action of $NaCl$ from the coke in forming silicates which are fusible at moderate temperatures. .

(4) The deposition of carbon around spots of iron which occur in the bricks, and which cause disintegration, especially in the upper part of the furnace.

The first two causes are undoubtedly extremely active in their respective zones, but the last two can safely be neglected if they exist at all.

The carbonization of the bricks below the fusion zone tends to protect them by offering to the corrosive effect of the cinder a substance which is neutral to both acids and bases. It is probable that no bricks could long withstand the furnace conditions without this protection. The carbonization of the lining appears to be most rapid under the influence of a hot, limey cinder, which is an additional reason for blowing in hot and limey, so that the protecting condition may be produced as early as possible. The carbonizing of the lining led to the suggestion of using carbon bricks, but they have never come into general use.

Tr. A. I. M. E., XXI., 112.

For the protection of the bosh walls, which is the most vulnerable part of the furnace, bosh-cooling plates are in almost universal use. They preserve the proper shape of the walls, thereby prolonging the blast, decreasing the fuel consumption and increasing the output.

Hot Spots.—With the introduction of automatic charging, the hot spot assumed a new importance. It was soon found that furnaces charged with a self-dumping skip usually developed a hot spot on the shell about 20 feet above the top of the bosh and usually on the side opposite the skipway, within a few weeks after blowing in. It is generally assumed that this is the effect of the channeling of the gases through the lumpy portion of the materials which are thrown to that side. A similar effect will result from an improperly proportioned bell, since the lumps always roll to the lowest level, and the fines remain behind, forming peaks or ridges immediately below the point of discharge As a result, automatic chargers have been pretty uniformly supplemented by stock distributors. The fact that they have generally succeeded in mitigating this difficulty shows pretty conclusively that it is due to the irregular distribution of the furnace action. Since the location of the trouble is generally far above the usual zone of fusion, it is evident that if the action is due to fluxing of the lining, it must result in a very easily fused silicate, which points to action by the alkalis. This theory is supported

Tr. A. I. M. E., XVI., p. 149.

Tr. A. I. M. E., XXXV, p. 224.

by the fact that the level corresponds to that of the decomposition of the cyanides and the formation of alkaline carbonates.

Breakouts. —A breakout resulting from destruction of the lining, may occur at any point below the limits of fusion, but does not assume a very serious significance unless it occurs below the surface of the metal lying molten in the hearth. The emission of iron or cinder around the coolers is not followed by any considerable body of matter, and may be readily checked by chilling with water from a hose. If the breakout occurs at the tapping-hole, the iron should be directed into the pig-beds, or other suitable depressions, so that it can be readily broken and handled. Sometimes breakouts find their way through the foundations to some distance and rise through the floor around the furnace, where they must be relieved by promptly tapping the furnace. Breakouts are never due to the erosion of the hearth lining, because iron is not active in disintegrating brick. They can occur only when the brickwork becomes disrupted from any cause, such as expansion, thereby leaving a passage to the unprotected shell. They may be prevented by properly jacketing and binding the hearth wall, to prevent movement. Frequently they stop themselves by chilling in the cracked brickwork before they reach the jacket. Water-cooled jackets or heavy uncooled jackets having large thermal capacity will chill incipient breakouts, thereby checking them before they become dangerous.

OBSTRUCTIONS.

Several different types of furnace obstructions may be distinguished as occurring at different levels of the furnace and being due to different causes.

Pillaring. —Pillaring is a phenomenon peculiar to the hearth, and is the result of faulty blast distribution. If the penetration of the blast is insufficient, there is liable to be a conical pillar of cold stock extending up through the middle of the hearth and surrounded by an annular column of activity. The presence of the pillar has been proven in some instances by thrusting a bar across the hearth from tuyere to tuyere. The obstruction could be felt, and the bar, when withdrawn, was red at each end with a black middle section. An increased blast volume or a decreased tuyere area would give better blast penetration to such a hearth.

Pillars of dead material may accumulate around the hearth between tuyeres, especially if the tuyeres are far apart and the blast enters under good pressure. Such pillars are not necessarily detrimental. They may be reduced in size by increasing the number of tuyeres.

Scaffolds. —Obstructions which have their origin above the hearth and depend more or less upon the furnace walls for support are known under the somewhat general term of " scaffold." A scaffolded furnace is sometimes said to be " hanging," an expression which is derived from the fact that charges refuse to descend regularly and properly. Scaffolds are very prone to occur on the bosh of the furnace, where they have their origin in incrustations due to a contraction of the zone of fusion. They occur also in the zone of reduction, where they appear to be due to the wedging resulting from excessive carbon deposition. Such conditions result in frequent slipping of the stock, causing ir-—regular quality of iron, due probably to the accession of oxygen brought in by the sudden precipitation to the hearth of incompletely prepared materials. This irregularity is far reaching in its effects, and is frequently the cause of much of the " seconds " in the rail mill.

Incrustations occur at the top of the fusion zone. If for any cause the zone of fusion undergoes contraction, the pasty, partially melted material becomes incrusted and adheres together and to the walls, thereby impeding, and sometimes preventing descent. The partially fused condition of the incrustation may offer very great resistance also to the passage of the gases. The contraction of the zone of fusion may be brought about by several causes, such as decrease of fuel in the charge, increase of refractoriness of the slag, or increase of moisture in the blast, on account of which the quantity of heat developed may be less than the requirements. It is sometimes caused, also, by too high blast temperature. This naturally brings more heat into the hearth, but may also cause a contraction of the zone of fusion. This paradox is best accounted for by the fact that the activity of oxygen toward carbon increases as the temperature rises. The combustion takes place sooner, in consequence, and is proportionally more intense and less extended in area.

The best indication that a furnace is scaffolded is its refusal to take charges or the indication of the gauges that the stock does not sink, or else sinks more rapidly on one side. The blast pressure usually goes up, and the gas becomes thin, hot and scanty. The gas flames and the fumes become less dense. If the obstruction is due to incrustation, a portion of cold blast will drive the zone of fusion higher and will tend to soften the obstruction. Then slacking the engines and opening the relief valve will frequently let the obstruction descend It is advisable to give the furnace blast again as soon as the obstruction is loosened, so that the upward pressure may form a cushion to prevent splashing the tuyeres with cinder and iron. It is best also to open the bell to relieve the sudden rush of gas displaced by the falling mass. This method is not always successful at first, but a second or third trial may bring results

Repeated hangs may be caused by too much limestone, in which case they may be relieved by reducing the quantity of stone or by charging sand or sandy scrap over the scaffold. Scrap and coke may be charged on an obstruction, but stone should always be avoided. More persistent hangs may demand more strenuous measures. Various methods have had their advocates. A method considerably used formerly consisted of removing the cinder notch and cooler and by means of a full blast, blowing out through the notch all of the stock below the obstruction, thus leaving it unsupported till it falls under the weight of the superimposed material. An obstruction which adheres pertinaciously may be removed by a charge of dynamite or by local combustion. A hole may be drilled through the furnace wall back of the obstruction and a tuyere temporarily introduced. If the charge at the point contains insufficient fuel, an oil blowpipe may be substituted. An obstruction which has formed an arch across the furnace and has interrupted all passage of gas or stock may be broken up by exploding dynamite above it

Tr. A I M E, IX , p 60.

Tr A I M E., XIII. p 670.

Repeated obstructions of the same nature show faulty management and may be due to too heavy burden or improper fluxing. Scaffolds may be sometimes traced to faulty furnace design, such as improper bosh angle or improper size of bell. In case of faulty design, the remedy lies only in remodeling the furnace.

The form of obstruction, due to **wedging** of the stock, frequently occurs in the zone of carbon deposition in the top of the furnace. It has been shown by Laudig that the weight of carbon deposited on some ores may exceed the weight of oxygen extracted by almost 50 per cent., and even exceed the whole ore in volume. In his experiments, the average of 33 ores of various types and localities showed that the usual deposition of carbon amounted to about 60 per cent. of the oxygen extracted. On the assumption that 90 per cent. of the oxygen of the ore is extracted, in the zone of carbon deposition, we would expect an average deposition of 23 pounds of carbon for each 100 pounds of pig, or about 170 pounds of ore. As the specific gravity of the ore is several times greater than that of the deposited carbon, it appears that such a deposit of carbon would nearly double the volume of the ore. Such a large increase of volume is out of proportion to the usual batter of furnaces at that depth, and the inevitable result is wedging. The carbon also fills the interstices between the pieces of stock, and prevents the proper flow of the gases.

When the stock becomes wedged so tightly that it can no longer descend, the furnace is said to "hang." The unwedged stock beneath continues to settle and ultimately the unsupported bridge falls. This is known as a "slip" and is usually accompanied by an "explosion" of greater or less violence, which frequently ejects considerable quantities of stock from the furnace and occasionally displaces the bell and hopper. These so-called explosions are not of the nature of the usual phenomena called by that name, but are generally somewhat prolonged blows, as if they represented merely relief from an accumulation of tolerably moderate pressure, rather than an explosion proper. A genuine explosion sometimes occurs in the top of the furnace, when, for any reason, air is drawn in in sufficient quantity to make an explosive mixture with the furnace gases. Such an explosion would naturally occur in the open space above the stock, and while likely to damage the furnace, could hardly eject stock which lies below it. Yet stock is generally ejected. Sometimes as much as three charges have been thrown out, and there is a case on record where a charcoal furnace was practically emptied by this action. It has been suggested, therefore, that the explosion might be due to sudden reduction of the ore by finely divided carbon, which thereby generates rapidly a large volume of gas. It is probable, however, that a true explosion well down in the furnace would burst the shell before it could eject such a heavy mass of superincumbent materials The suggestion that the effect is due to the sudden displacement of the gases on account of the falling mass is open to the objection that some of the smallest slips show the greatest power. Moreover, simple movement of a confined gas does not necessarily develop an increased volume or pressure. The true explanation is probably embodied in the theory that the gases are trapped until the presence of the blast is sufficient to force a passage and some of the stock is carried out by the momentum.

It can easily be shown that three charges of stock will weigh, when charged, approximately 100,000 pounds, and in a furnace having a 15 foot stock line, will occupy a depth of about 10 feet. Through loss of C, O_2 and CO_2, this stock may not weigh more than 90,000 pounds. The upward pressure of the gases at that point when a blast pressure of 15 pounds is consumed in driving the gases through 75 feet of stock will evidently be approximately 2 pounds per square inch, which exerts a total upward pressure of about 51,000 pounds. It is evident, therefore, that even under normal conditions, the lifting force of the gases at 10 feet depth is equal to more than half the weight of the superincumbent charges.

If, then, the stock beneath continues to settle until a gap of 10 feet exists, there will be a column of materials only 55 feet high for the blast to traverse. If this resistance uses up 11 pounds of the initial pressure, there will be left 4 pounds pressure at the depth of 10 feet, which is double that before, and more than enough to lift bodily the whole quantity of stock above it. Consequently, it breaks through the weakest point in the wedge, and the suddenly released pressure carries some of the stock with it out of the furnace and lets the rest fall within. This assumption

Blast Furnace.

is based upon continuance of normal pressure in the engine room. If the engine speed has remained constant, and the pressure has risen in consequence, the moment when the accumulated pressure will force a passage will arrive much sooner.

The frequency of these so-called **top-explosions** was greatly increased by the use of high percentages of Mesaba ore. The trouble was naturally attributed to the fine state of division of the ore, since that was its most marked peculiarity. However, it is observed that finely divided magnetic concentrates do not cause slips and explosions in anything like the way that Mesaba ores do. When it is recalled that the Mesaba ores deposit by far the most carbon of any ores known, while magnetites deposit little if any, this fact confirms the opinion that the difficulty is due not so much to fineness, *per se,* as to resistance of deposited carbon.

Prevention of Wedging.—The wedging due to carbon deposition can be prevented by avoiding ores that are active in separating carbon. If such ores must be used, means for prevention of this pernicious activity should be adopted. It has been observed Iron Age,
May 5, 1904,
p. 6. that the use of high percentages of limestone has this effect. For example, a furnace which gave trouble with 21 per cent. of stone, ran very smoothly on 27 per cent. The most obvious explanation of this fact is that the solvent power of the additional 30 per cent. of CO_2 which was evolved, was sufficient to dissolve enough deposited carbon to enable the gases to keep an open passageway, whereas the lesser amount could not. Again, it has been observed that stone crushed to small sizes is not as efficient in this respect as large stone. A furnace which ran smoothly on 30 per cent. uncrushed stone, immediately gave trouble when the stone was crushed to pass a 4 inch ring. It is not improbable that the CO_2 was evolved from the smaller pieces too high in the furnace to exert its full solvent power.

CHILLED HEARTH.

A chilled hearth is probably the most serious disorder that can befall a furnace. It may have its origin in any cause that does not leave enough heat in the hearth to keep its contents fluid. This condition may result from an insufficient supply of fuel, or from a suddenly increased demand upon the regular supply, such as leaking water coolers or excessive moisture in the blast; or it may result from the sudden precipitation into the hearth of a mass of material from a colder part of the furnace. When the hearth chills, the iron and cinder become solid either wholly or in part, which makes tapping in the usual way impossible. The only remedy is to raise the temperature of the hearth to the melting point as soon as possible.

Detection of Chill.—When chilling is due to a too heavy burden, or to too moist blast, the effect is gradual and may be foreseen. The signs are a cold cinder and cold iron, dark tuyeres, chilled cinder notch and tuyeres. By promptly charging a coke blank, the condition may be corrected before very serious results occur.

When the chilling is due to copious water leaking, or to a slip, the effect may be so sudden as to preclude the possibility of prevention. A chill caused by a small amount of water leaking into the front of the furnace will give rise to a hard tapping hole. If the leak is not large, the hardness may be evident only after the blast has been shut off for a while. The chilling may not be serious, simply lengthening the time necessary to open the tapping hole. If it is impossible to open the hole, preparation must be made to tap iron through the cinder notch. The system of coolers must be removed and a temporary runner of bricks and clay constructed.

Remedy for Chill.—After a temporary system of casting through the cinder notch has been established, the tapping hole may be opened up by some auxiliary system of melting. The time-honored method is by means of the oil blow pipe, which is a device for spraying inflammable oil by means of an air blast. A 3 inch pipe, tapped from the bustle pipe or tuyere stock by means of a flexible connection, conveys the air, and a ⅜ inch pipe led into the 3 inch pipe through a reducing flange, is connected with an elevated supply of oil. The blast breaks up the drops of oil into a fine spray, which burns with the oxygen of the blast, producing a very high temperature, which is sufficient to melt the material chilled in the tapping hole.

The same effect can be produced by means of the electric arc. By connecting the positive pole of a generator with the hearth

Tr. A. I. M. E., XV., p. 417.

Ibid, XVI., p. 779.

Tr A I M. E.,
XXXI, p. 626.

jacket and applying the negative pole connection in the shape of
a large carbon to the tapping hole, an arc may be established that
will readily melt obstructions. To obtain satisfactory results,
however, the current should not be less than 400 amperes, at 220
volts pressure. It is still better to have 1,000 amperes, at 110
volts, when the action is said to be more rapid than the oil blow
pipe. It is necessary, however, to protect the eyes with heavily
smoked glasses, as the rays from the arc produce after effects
which incapacitate the unprotected beholder for a day or two.

A still more recent method, known as the Menne process,
consists of burning the iron itself by means of compressed
oxygen. The apparatus consists of an ordinary oxyhydrogen or
Iron Age,
May 17, 1906.
hydrocarbon blowpipe with a long nozzle, by which the temperature of the point to be attacked is raised to incandescence. The
supply of fuel is then shut off, and the oxygen alone played
against the heated metal at a pressure of 30 atmospheres. The
heat developed by the oxidation of the metallic iron and the metalloids is sufficient to render the resulting oxides into a thoroughly
fluid condition and the pressure of the oxygen blast keeps the
hole clear. A higher temperature can be obtained from the burning metal than from the hydrogen, as the volume is so much less
that the heat is better concentrated. It is said that this method
can penetrate nearly a foot of metal per minute. It is manifestly
useless, however, when slag forms the obstruction, since its very
existence depends on an oxidizable obstruction.

When a large quantity of scaffold material falls into the
hearth, it not only chills the tapping hole, but frequently the
cinder notch as well. At the same time, it usually forces molten
iron and cinder up around and into the tuyeres, where they chill.
It is not uncommon, therefore, to have every opening into the
lower part of the furnace chilled up tight. Under such circumstances it is impossible to keep the blast on the furnace, since
the tuyeres are closed. It is necessary then to open them by an
auxiliary means of melting, and get the blast on as soon as
possible. This is best done by attacking two or three points at
once, such as the cinder notch and an adjacent tuyere or both
adjacent tuyeres. This will soon create ingress for blast at two
tuyeres and provide an outlet for molten products as fast as made.

The other tuyeres may be recovered in succession while the cinder-notch serves as outlet for both cinder and iron. Oil fed into the belly pipes will act as a blow pipe until good fuel again appears at the tuyeres. When melting has been restored, a coke blank should be charged, and then an outlet through the tapping-hole made by the blow pipe. By means of internal and external combustion the hearth can be gradually restored to normal condition.

CHAPTER VII.

HINTS ON DESIGN AND EQUIPMENT.
FURNACE DESIGN.

In the construction of plants for producing pig iron, experience has shown that practically any expense is justifiable that leads to material economy of operation. There are three cardinal points in economical operation that may depend directly upon plant design, first, large output, second, low fuel consumption, and third, low cost of handling materials and products.

The output of a furnace plant may be considered as dependent upon two chief factors—viz., the size of the furnace hearth and the capacity of the blowing equipment.

The fuel consumption will depend upon the furnace lines and the heating capacity of the stoves.

The cost of handling materials and products will depend upon the design of the stock handling arrangements and the systems of handling iron and slag.

Size of Hearth.—The keynote of every furnace plant is its output. The dominating factor in the question of output is the size of the furnace hearth. The rate of smelting must always be in proportion to the rate of fuel combustion. The rate of combustion under given conditions is tolerably constant per square foot of hearth. The output must, therefore, depend largely upon the hearth area. It is true that the nature of the ore, the kind of fuel and the quality of product all have a bearing upon the rate of smelting; but with average mixtures of ores and average quality of coke and usual forms of pig, every well-working hearth should consume at least 6000 pounds of coke per square foot in 24 hours. With this assumption as a basis, the hearth area for any output at a given fuel consumption may be figured. For example, a furnace to make 400 tons of iron in 24 hours in a fuel consumption of 2240 pounds, would require a hearth area of $\frac{400 \times 2240}{600} =$ 150 square feet, or a diameter of about 14 feet.

For unusual ore or fuel conditions, it is necessary to acquire experience as a guide to the especial needs. Charcoal, because of the large surface which it presents to the blast, permits a more rapid rate of combustion than coke, and hence a higher duty per square foot of hearth area. On the other hand, the substitution of anthracite coal for coke would not permit more than half the full estimated production of the hearth.

The hearth area should be adapted to the kind of iron to be made. Small hearths are better suited to making foundry irons, because they concentrate the heat and reducing conditions and thereby produce higher silicon content. At the same time, however, the higher fuel ratio demanded by foundry irons decreases the duty per square foot of hearth area and a lower output follows. Inst. Jour, 1901, I, p. 158

The crucible capacity should be about 3 cubic feet per ton of pig. For a 14 foot hearth, this means a depth of about 8 feet. Since each ton of iron occupies 5 cubic feet of space, it follows that if the furnace is to be tapped every six hours, the well below the cinder notch must have a depth of at least 3½ feet. If tapping occurs oftener, a lesser depth may suffice, but is not advisable. Tr. A. I. M. E., XXXIV., p. 608.

Bosh.—When the size of the furnace hearth has been decided upon, every other dimension should be brought into harmonious proportion. The most successful of the modern furnaces have bosh areas two to two and a half times that of the hearth. A hearth of 14 feet diameter would therefore require a bosh area of 300 to 375 square feet, which corresponds to diameters of 20 to 22 feet. If we take 21 feet diameter as the size of the bosh, a bosh angle of 75 degrees from the horizontal would bring the bosh at a height of about 13 feet above the top of the crucible.

The bosh angles best suited to different materials can be determined only by experience. The usual slope to-day is from 70 to 80 degrees, with best results at about 75 degrees. Less than 70 degrees slope is so flat that it permits accumulations which slip periodically, causing irregular work. More than 80 degrees is likely to bring the bosh above the zone of fusion and permit hanging.

Shaft.—A stockline diameter about equal to that of the hearth is ample for all purposes. The location of the stockline is

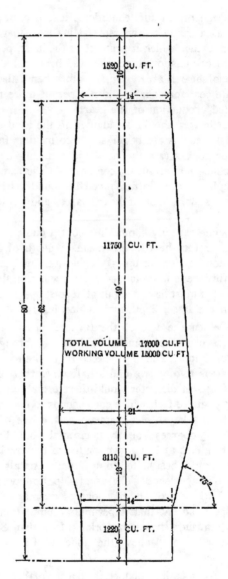

Lines of a Modern Blast Furnace.

determined by the position and drop of the bell. Usually the space occupied by the hopper, plus the drop of the bell, and clearance below it, will leave at least 10 feet of dead space between the charging platform and the stockline. For an 80 foot furnace this

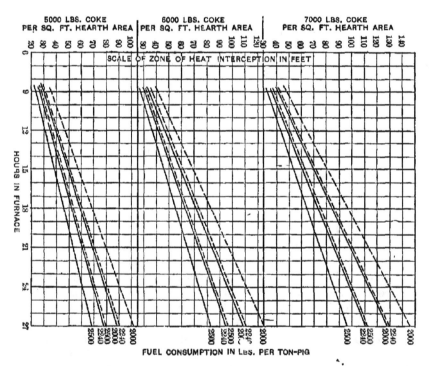

Diagram Showing Relation Between the Fuel Consumption and Rate of Combustion on the One Hand and the Length of Heat Intercepting Zones and Rate of Travel of Stock on the Other, at Various Fuel Consumptions and Bosh Ratios.
(Full Line = Bosh 2½ Times Hearth Area.)
(Broken Line = Bosh 2 Times Hearth Area.)

would leave 49 feet as the distance from the bosh to the stockline. The taper from 21 to 14 feet gives the inwalls a total batter in that distance of 42 inches on every side, which is equal to 0.85 inches per foot. The batter varies in general practice from ½ to 1 inch per foot. This gives a total volume of 17,600 cubic feet or

44 cubic feet per ton of iron in 24 hours With Lake Superior ores the latter volume generally ranges from 30 to 50 cubic feet.

TABLE SHOWING AIR REQUIREMENT, OUTPUT AND LENGTH OF ZONE
OF HEAT INTERCEPTION FOR VARIOUS FUEL CONSUMPTIONS
AND BOSH RATIOS

Fuel con-sumption	Pounds coke burned per sq. ft. of hearth in 24 hours.	Cubic feet air per sq. ft. of hearth per minute	Pounds of iron per sq ft of hearth in 24 hours	Gross tons of iron per sq ft in 24 hours	Bosh area 2½ times hearth area. Time of stock in furnace			Bosh area double hearth area. Time of stock in furnace		
					12 h Feet.	15 h Feet.	18 h Feet	12 h Feet.	15 h. Feet.	18 h. Feet
2,000 lb per ton	5.000	191	5,600	2.5	39	49	59	47	58	69
	6,000	229	6.720	3.0	47	59	71	55	69	83
	7,000	267	7,840	3.5	55	69	83	64	80	96
2,240 lb per ton	5.000	191	5,000	2.232	34.5	43.2	51.8	40.3	50.4	60.5
	6,000	229	6,000	2 679	42.0	52.0	62.0	49.0	60.5	73.0
	7,000	267	7,000	3.125	48.3	60.4	72.5	56.5	70.6	84.7
2,500 lb. per ton	5.000	191	4,480	2.0	30.4	38.0	47.6	35.5	44.3	53.1
	6,000	229	5,376	2.4	36.4	45.5	54.6	42.5	53.1	63.7
	7,000	267	6,272	2.8	42.6	53.2	63.9	49.6	62.0	74.4

FURNACE CONSTRUCTION.

Foundations.—The foundation of a blast furnace cannot be too solid It should reach down to bedrock or hardpan, and below the general level of the yard may consist of concrete. Above the yard level, however, the construction should be of brick. For convenience of handling the product in ladles or cars running on the general yard level, it is always desirable to have the hearth level at least 10 feet above the yard level. The construction from the yard level to the hearth level is best made of firebricks, although red bricks may be substituted outside the circle marked by the hearth jacket. The space between the foundations and the cast house wall should be filled with loose material in order to check breakouts.

Columns.—The columns which support the mantle are placed upon the foundations at the hearth level. The columns may be cast iron or may be built up of structural steel. It is difficult to get cast columns with metal of uniform thickness when more than 18 feet long, and as recent columns are sometimes 24 feet, there is a tendency to use built up columns. It is convenient to have half as many columns as tuyeres, so that the tuyeres may

TABLE SHOWING THE CONDITIONS WHICH NORMALLY RESULT FROM VARIATIONS IN HEARTH DIAMETERS.

(a) Hearth diameter in feet, and no. tuyeres.	(b) Hearth area in square feet.	(c) Air needed per sq. ft of hearth consuming 6000 lb coke in 24 hours. Cu. ft.	(d) Air needed by furnace per minute. Cu. ft.	(e) Number of revolutions of engine with air cylinder 84 x 60	(f) Total heating surface required in stoves combined Square feet.	(g) Pig produced and coke consumed at fuel consumption of 2240 lb. Tons.	(h) Boiler capacity required at 8 per ton coke H.P.	(i) Diameter of tuyeres.—Inches.	(j) Minimum diameter of bustle pipe.—Inches.	(k) Bosh diameter.—Feet.
8	50.3	230	11,600	31.5	58,000	135	1,100	4½	12½	12½
9	63.6	230	14,600	39.5	73,000	170	1,400	4½	14	14¼
10	78.5	230	18,000	48.5	90,000	210	1,700	5	15½	16
11	95.0	230	21,800	59.0	109,000	250	2,000	5½	17	17¾
12	113.0	230	26,000	70.0	130,000	300	2,400	5½	18½	18¾
13	132.7	230	30,500	82.0	152,500	350	2,800	5½	20	20½
14	154.0	230	35,400	95.0	177,000	410	3,300	6	21½	22
15	176.7	230	40,600	110.0	203,000	475	3,800	6	23	23¾
16	201.0	230	46,200	125.0	231,000	540	4,300	6	24½	25¾
17	227.0	230	52,200	141.0	261,000	610	4,900	6½	26	27
Assumed.	$0.7854\times a^2$	230	$\dfrac{6000\times55}{1440}$ $(b)\times(c)$	$\dfrac{(d)}{10000}\times27$	$5\times(d)$	$\dfrac{6000}{2240}\times(b)$	$8\times(g)$	$8\times\sqrt{\dfrac{(d)}{100\times(a)}}-0.7854$	$\sqrt{\dfrac{(d)}{100}}-0.7854$	$\sqrt{2\frac{1}{2}\times(b)}-0.7854$

253

be placed symmetrically, two between every pair of columns. The columns support the mantle and are bolted at the bottom to heavy cast iron base plates, which rest on the foundations.

Hearth.—The crucible jacket may be iron or steel castings made in segments and bolted together, or it may consist of riveted steel plate, 1 inch or more in thickness, reinforced by heavy bands. The latter construction is more rigid and also cheaper than the cast segments. The jacket should be set well into the foundation brickwork, at least 4 feet below the hearth level, and should be cylindrical in shape. Conical jackets tend to move upward during expansion, and any movement is likely to permit the formation of cracks in the brickwork which may lead to breakouts. An opening in the front of the hearth jacket at least 12 inches wide and 30 inches high should be left at the hearth level for the tapping hole. It should be surrounded by a Z bar collar to shed water.

The crucible jacket is generally cooled to prevent breakouts, and the system of cooling is modified by the style of jacket. Cast jackets are sometimes so heavy that the mass of metal acts as a chill. Generally, however, they are cooled by water flowing in wrought iron pipes enclosed in them when they are cast. Sometimes external gutters filled with flowing water are used. Riveted plate jackets may be cooled by sprays of water directed against the external surface, or by internal rows of vertical wrought iron pipes, laid either against the jacket or incased in cast iron. Surface cooling with water under gravity is preferable to internal pipes, which requires water under pressure, because the latter is difficult to manage when a breakout cuts the pipe.

Cinder Notch.—The cinder notch usually comes at about the top of the crucible jacket to which the cinder runner is bolted. The cinder notch usually consists of three pieces, although the intermediate ring is sometimes omitted. The water-cooled parts are usually copper or bronze, though the outer cooler is often cast iron containing a coil of wrought iron pipe. The latter is preferable whenever it is necessary to cast through the cinder notch, as the cast iron is less quickly destroyed by molten iron than copper or bronze.

Furnace Level.—The working level around the furnace is

usually 2 or 3 feet above the hearth level, except at the front, where it is lower to give access to the tapping hole. With surface cooling of the hearth jacket, an open drain 4 inches wide called the " well " is provided in the brickwork around the crucible walls, to allow the cooling water to reach the bottom of the jacket. This drain should be kept full of gravel or other porous filling, which will allow free drainage. Formerly it was customary to have wells 1 to 2 feet wide kept full of water, which in case of break-outs caused disastrous explosions and permitted the wells to fill with iron.

Tuyeres.—The crucible wall is pierced for tuyeres, so that their centres are 18 to 24 inches below its top, which allows suffi-cient space for the tuyere-breast casting below the flare of the bosh wall. In the case under supposition, this would leave about 3 feet between the cinder notch and the line of the tuyeres. Since the bottom of the tuyeres marks the limit to which cinder should rise, this space should be as high as possible.

The number of tuyeres used for a given size of hearth varies widely, although there is a tendency to come to the uniform rule of a tuyere for each foot of hearth diameter. This rule gives a uniform distance between tuyere centers of almost exactly 3 feet.

The tuyere should not be so small that it will throttle the blast, nor so large that the blast will lack penetration. The best results have been attained when each square inch of tuyere opening passes about 100 cubic feet of piston displacement per minute. In order to pass 35,000 cubic feet of air per minute at this rate, 350 square inches will be required; 14 tuyeres, therefore, would need a diameter of 6 inches each in the clear.

Tuyeres are sometimes made in three pieces like the cinder notch, but as a rule the tuyere and cooler alone compose the system. The average projection of the tuyere beyond the nose of the cooler is about 6 inches, which seems to give the best results. Less projection allows the combustion to attack the walls above the tuyeres. More projection simply reduces the working area of the hearth. Attempts to distribute the blast horizontally by special shapes of tuyere nose causes cutting of walls beside the tuyeres.

Copper tuyeres cost somewhat more than bronze, but, owing

to their better conductivity, they resist burning better and stand
abrasion quite as well. They should be fed at the bottom by a 1¼
inch pipe and the outlet should be reduced to ¾ inch or less.
The outlet pipe should be at the top and should extend well
toward the nose. This secures a better circulation at low pressures. Fifteen to twenty-five pounds is ample. A small opening
in the outlet pipe at the top of the tuvere base allows for escape
of any trapped air.

Bosh Construction.—The bosh construction is, in the majority of cases, brick with bronze cooling plates. Since they offer
but little resistance to outward pressure, they must be reinforced
by heavy iron bands. Bosh plates of the Scott type with arched
top, and sides tapered and nose curved to coincide with
the curvature of the wall are most used. They extend the full depth of the wall, and are kept filled
with water by a positive circulation. As a rule, a
number are connected in series, so that the discharge from
one becomes the supply of the next. Usually the temperature of—
the water does not rise more than 10 to 20 degrees F. in passing
through a plate, so that 125 degrees F. is a perfectly safe limit
for inlet water. The space between plates horizontally is usually
one brick, 4½ inches, so that the proportion of the plates to the
circumference is 75 to 85 per cent., according to the size of plate
used. Small sizes are preferred. Usually they are spaced vertically 12 to 24 inches apart. Hence a 21 foot bosh 13 feet deep
will require about 350 small plates. Above the bosh, the lower
part of the inwall is generally protected by two or three rows
of cast iron cooling plates or coils of pipe set in the brickwork.

Equally good for small furnaces, and probably as good for
any size, is the steel plate shell construction with surface cooling.
It is a stronger form of structure than bricks reinforced by bands,
and much less expensive than a multitude of bronze bosh plates.
Furthermore, only about one-third the quantity of bricks is needed
for lining, since a 9 inch or 13 inch wall suffices. A segmental
cast bosh jacket with troughs or pipe circulation is still in use in
some districts.

Furnace Shell.—The shell of the furnace shaft is always of
riveted plate-construction and is supported by a mantle which is

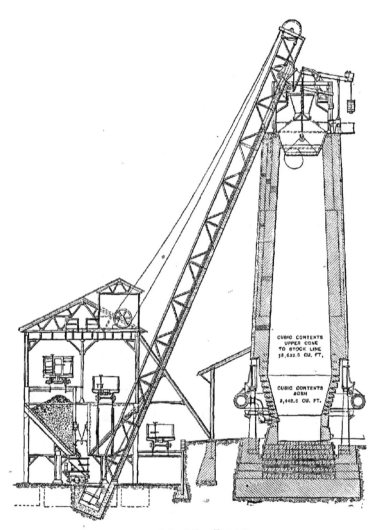

Section of the Eliza Furnace.

bolted to the top of the columns. The mantle is sometimes a segmental cast iron ring, but now is more usually built up of plates and shapes. The furnace shell usually has a taper conformable to the batter of the inwalls. Near the top it is pierced for the downtake openings, and frequently also for a row of explosion doors, whose combined area may be one-half the area at the stockline. The explosion doors, however, do not appear to be necessary as the so-called "top explosions" are not due to increase of the volume of the gases, but rather to their inertia of motion on sudden release from obstructions. In order that the gases may not carry away too much ore as flue dirt their rate of escape through the downtake should not exceed 32 feet per second. For a gas volume of 1460 cubic feet per second at 450 degrees F., resulting from 35,000 cubic feet of blast per minute, a downtake at least 7 feet in the clear is necessary. This would mean one opening of 7 feet in diameter, or two of 5 feet, or three of 4 feet, or four of 3½ feet.

Furnace Top.—The furnace top is closed with a bell and hopper so designed that the bell has 2 to 2½ feet clearance all round it at the stockline. The hopper should have an available capacity sufficient to hold all of the coke needed for one charge. The use of a second bell above acts as a seal and saves the waste of gas incident to dropping the charge. The systems of distribution, which provide for the rotation of stock, appear to give best results.

For filling the furnace two balanced skips travelling on parallel tracks work best. The size of skip should be in accordance with the character of stock and rate of travel of the skip. Ordinarily the stock necessary to make 100 pounds of pig occupies 5 to 5½ cubic feet of space. A furnace which makes 400 tons of pig iron in a day must make 640 pounds per minute and hence use 30 to 35 cubic feet of stock in that period.

Stock House.—In order to supply the skips with stock promptly and economically it is necessary to have proper stock house arrangements. The most approved method consists of a series of bins for ore and stone, ranged generally at right angles to the skipway, and served by an electric larry, which weighs the stock and conveys it to the skip. The coke bins are placed on either side of the skip so that the coke may be drawn directly into the skip and charged by volume instead of by weight. The arrangements for the ore and stone consist either of a series of bins, surmounted by an elevated track, or a floor provided with a row of chutes leading to a tunnel beneath. The latter arrangement is less subject to freezing and is equally adapted to hand charging.

For storing ore for future use, it has been customary to use continguous space served by devices for economical rehandling of the stock in bringing it to the stock house. There is a recent tendency, however, to put the storage yard near the hoist and have it all served by the larry, thereby entirely obviating the necessity of rehandling.

The space occupied by stock when thrown loosely in bins is usually about as follows:

Lake ores . . . .150 pounds occupy 1 cubic foot space.
Stone   . . . . .100 pounds occupy 1 cubic foot space.
Coke   . . . . 28 pounds occupy 1 cubic foot space

The bin capacity required by a 400 ton furnace will evidently be as follows:

Ore $\dfrac{400 \times 2,240 \times 1.7}{150} = 10,000$ cubic feet per 24 hours, or 25 cu. ft. per ton pig

Stone $\dfrac{400 \times 1,120}{100} = 4,480$ cubic feet per 24 hours, or 11 cubic feet per ton pig

Coke . $\dfrac{400 \times 2,240}{28} = 32,000$ cubic feet per 24 hours, or 80 cubic feet per ton pig

Furnace Linings.—A considerable factor in the economy of operation of a blast furnace is the lengths of the blasts. The lengths of the blasts in turn are largely dependent upon the quality of the furnace lining

Furnace linings are always made of firebricks. The character of the bricks needed differs in different parts of the furnace As a rule three kinds are used The most refractory are needed for the hearth and bosh walls. Those forming the inwalls should be dense to stand the wear of descending materials and resist the action of the gases. The top bricks which receive the shock of the material as it slides off the bell should offer infinite resistance to abrasion.

All firebricks have **fireclay** as their basis. Fireclay is a hydrous

silicate of alumina, resulting from the decomposition of the feldspars which occur in granites, porphyries and other igneous rocks.
During the decomposition, the feldspars break up into silicates of
alumina and the alkalis. The latter, being readily soluble, are
leached out by circulating underground waters, leaving behind the
clay, mingled with other components of the parent rock, such as
quartz, mica and often some undecomposed feldspar. Pure clays
have the following approximate composition:

$$SiO_2 \dots \dots \dots \dots \dots \dots \dots \dots 47 \text{ per cent.}$$
$$Al_2O_3 \dots \dots \dots \dots \dots \dots \dots 40 \text{ per cent.}$$
$$H_2O \dots \dots \dots \dots \dots \dots \dots \dots 13 \text{ per cent.}$$

and any considerable variation from these proportions indicates
impurities which are not essential to the substance.

Fireclay has two properties which render it valuable in the
manufacture of refractories, namely, plasticity and refractoriness.
Plasticity is a quality inherent in clay and peculiar to it. It is
essential to the shaping and the persistance of shape of refractory
articles. It is impaired by the presence of any non-plastic substances, such as quartz, mica, feldspar, limestone, oxides of iron,
etc , and is also affected by high temperatures, excessive pressure
or any cause which decreases the normal proportion of combined
water.

The quality of **refractoriness** depends upon both the chemical
composition and the physical condition. An increase in percentage
of Al_2O_3 or of both SiO_2 and Al_2O_3 as in calcination, tends to make
clay more refractory. The coarser the particles and the less in-
timately mixed, the less ready the fusibility. By mixing calcined
and uncalcined clays and non-plastic refractory materials of various degrees of coarseness, bricks of almost any degree of density
and refractoriness may be obtained. Excellence in firebricks depends upon several factors, especially proper grinding, bonding
and burning.

Hearth and bosh bricks should be made of the most refractory
clays without much bonding or excessive burning, since they are
subjected to heat only. The coarser the material the less readily
will it be fused. **Inwall bricks** should be more dense than bosh
bricks and hence should be more finely ground and more thoroughly burned. Fine grinding and burning at a temperature of

2600 degrees F. usually insures sufficient density. Hearth, bosh and inwall bricks should contain less than 2½ per cent. oxide of iron. **Top bricks** may contain more plastic clay than inwall bricks. They should also be finely ground and well burned. They are less refractory, but this is partially compensated by their becoming vitrified during burning. Hardness and density are more important than refractoriness. Well burned bricks will ring when struck.

In selecting bricks for lining a furnace, it is important to choose those which are suited to the requirements. Neglect of this precaution is a leading cause of unsatisfactory results. Soft, porous, refractory bricks should be confined to the hotter parts of the furnace, and never used where materials are still solid. Hard, strong bricks, which contain a high percentage of plastic clay are seldom refractory, and therefore should be used for no higher temperatures than that of the inwalls. Bricks should be uniform in size and regular in shape to insure good joints without the use of much mortar.

In laying the bricks of a furnace lining it is usual to leave a space between the shell and the lining for expansion. This space may be 3 to 4 inches wide, and is usually filled in with yielding material such as slagwool, granulated slag or loam and slag. Allowance should be made for vertical expansion, also. If the ironwork on top is not loose, at least 6 inches should be allowed in the lining of a furnace 80 feet in height.

In laying firebricks, lime mortar should never be used, as at high temperatures the CaO would attack the SiO_2, and Al_2O_3 of the clay. Firebricks should always be laid in a slurry of fireclay and water, too thin to be handled on a trowel. The bricks should be dipped in the fireclay and laid on the wall and hammered close to squeeze out all of the fireclay possible. Any excess beyond that necessary to fill the slight inequalities of the bricks may result in shrinkage cracks on drying. The top of each course should be slushed with a dipperful of the slurry.

STOVE DESIGN.

Since stoves are intended to heat the blast, it follows that they should be in proportion to the blast volume, and hence should bear a definite relation to the size of the furnace. They should be designed for a given volume of blast at a definite temperature. In attaining this end two factors must be considered—viz., the volume of bricks and the area of surface presented to the blast. Since it is undesirable that the blast temperature should vary materially, the weight of brick should be such that it can give out heat continuously for an hour without dropping more than 100 degrees F. in temperature. As the blast volume is usually three or four times that of the checker flue, it follows that a given particle of air is in the stove for only 15 or 20 seconds, and that therefore the heating surface must be ample to communicate its heat rapidly.

The duty of a stove per hour is evidently equal to the amount of heat carried in the blast per hour. In the case under consideration the furnace requires about 35,000 cubic feet of air per minute, which equals 2672 pounds, or about 160,300 pounds per hour. The heat carried in this quantity of blast may be found by the formula,

$$160,300 \, [0.2335 \, (t - t^1) + 0.0000208 \, (t^2 - t'^2)], \text{ where}$$
$$t = 1,200 \text{ degrees F}, \text{ the temperature of the blast, and}$$
$$t^1 = 100 \text{ degrees F., the temperature of the air}$$

This reduces to $160,300 \times 296 = 46,000,000$ B. T. U., which is the amount of heat to be furnished per hour by the brickwork. Allowance of 10 per cent. for losses, will bring this figure up to about 50,000,000 B. T. U. If this is to be accomplished with the loss of 100 degrees F., the quantity of bricks is easily calculated. The heat capacity of brick is about 0.2 B. T. U. per pound per degree F. In dropping 100 degrees F. they will yield

$$0.2 \times 100 = 20 \text{ B. T. U. per pound of brick.}$$

$$\frac{50,000,000}{20} = 2,500,000 \text{ pounds of brickwork.}$$

Assuming that a 9-inch fireclay brick weighs 8 pounds, and that 17 make a cubic foot of solid brickwork, then each cubic foot will weigh 135 pounds. Hence, 18,500 cubic feet or about 315,000 9-inch bricks are needed in each stove to furnish blast with a drop not exceeding 100 degrees F. per hour. This equals nine bricks per cubic foot of blast per minute, or about ½ cubic foot of brickwork per stove for each cubic foot of blast.

The cross section of a stove having a checkerwork 60 per cent. bricks and 40 per cent flues would represent a total volume equal to

$$\frac{18,500 \times 100}{60} = 31,000 \text{ cubic feet of checkers per stove.}$$

A stove 22 feet in diameter has an area of 380 square feet. Therefore the checkers must have a height of,

$$\frac{31,000}{380} = 82 \text{ feet.}$$

About 12 feet more are required for the dome, making a total height of the stove at least 95 feet. Since 1000 checker bricks will yield about 140 square feet of heating surface, the total heating surface per stove will approximate 44,000 square feet. For four stoves this will equal 176,000 square feet, or about 5 square feet per cubic foot of blast per minute, which practice shows to be ample.

— —

TABLE OF RELATIVE PROPORTIONS OF SOME OF THE CHIEF MAKES
OF HOT BLAST STOVES USED IN THIS COUNTRY.

	Combustion chamber.	Number passes.	Size of stoves, Feet.	Area of combustion chamber.—Sq. ft.	Area of flues, Square feet.	Ratio of combustion chamber to flue area. Per cent.	Total passes based on total stove section. Per cent.	Total heating surface. Square feet.	Equivalent of 9-inch bricks.	Area of heating surface per 1000 bricks.—Sq. ft.
Foote	Side.	2	22 x 100	43	116	2.7	41.8	54,000	352,000	153
Kennedy ....Central.	2	22 x 100	35	118	3.4	40.3	47,000	362,000	130
McClure .... Central.	3	22 x 100	33	124	3.7	41.3	52,000	395,000	132
Roberts, ...	Side	2	22 x 100	47	84	1.8	34.8	39,700	360,000	110

Stove Efficiency.—The quantity of gas required by the stoves and the maximum temperature obtainable may be determined from the above conditions. The gases which escape from the furnace have the following composition per pound:

CO .... 0.2384 pound.
CO_2 .... 0.1863 pound.
N_2 .... 0.5437 pound.

H_2 .... 0.0005 pound.
H_2O .... 0.0311 pound.
_____
1.0000 pound.

At a temperature of 450 degrees F., the quantity of heat existing in the gases per pound if cooled down to 60 degrees F. may be found as follows; when $t-t' = 390$, and $t^2-t'^2 = 198,900$;

B T U

CO 0.2384 [(0 2405 × 390) + (0.00002143 × 198,900)] = 0 2384 ×
    98 057 = ... 23.376
CO$_2$ 0 1863 [(0.187 × 390) + (0.000111 × 198,900)] = 0.1863 × 95 008 = 17 700
N$_2$ 0.5437 [(0.2405 × 390) + (0 00002143 × 198,900)] = 0 5437 ×
    98.057 = ...53.313
H$_2$ 0.0005 [(3.367 × 390) + (0 0003 × 198,900)] = 0 0005 × 1371 80 = 0.686
H$_2$O 0 0311 [(0 42 × 390) + (0 000185 × 198,900)] = 0 0311 × 200.496 = 6.255

Sensible heat available in gases per pound =.....................101.310

Allowing 10 per cent. to be lost by cooling through radiation from downtakes, distributing pipes and burners, we may realize about 90 B. T. U of sensible heat per pound of gas.

If the CO and H$_2$ burn to CO$_2$ and H$_2$O, respectively, in the presence of 50 per cent. excess of air, then the amount of heat developed per pound of gas will be,

B. T. U.
per pound

(0 2384 × 4,325) + (0.0005 × 51,700) = ... 1,057
Adding the sensible heat of the gases =............. 90

We have.. 1,147

as the total quantity of actual and potential heat per pound of— gases.

Assuming that the waste products of the combustion pass out of the stoves at a temperature of 600 degrees F , we may determine the amount of heat lost in the escaping gases. The final products of the combustion may be found as follows:

Composition of gases per pound air and changes they undergo.	Net air needed for combustion.		Products of combustion		
	O$_2$	N$_2$	CO$_2$.	N$_2$.	H$_2$O
0 2384 CO to CO$_2$ requires.....0.1362		0.4495	0.3746	0.4495	....
0.1863 CO$_2$	....	....	0.1863	....	..
0.5437 N$_2$	....	....	....	0 5437	..
0 0005 H$_2$ to H$_2$O requires...	0.0040	0.0132	....	0.0132	0.0045
0 0311 H$_2$O	...	....	....	....	0 0311
1 0000	0.1402	0 4627	0.5609	1.0064	0 0356

The air needed for combustion, 0 1402 + 0 4627 = 0 6029 pounds, whence 50 per cent. excess equals 0 3015 pounds.
The total available heat carried away at 600 degrees F may now be found as follows: When $t - t^1 = 540$ and $t^2 - t'^2 = 356,400$.

B. T. U

CO$_2$ 0 5609 [(0.187 × 540) + (0.000111 × 356,400)] = 0 5609 × 140.54 = 78 83
N$_2$ 1 0064 [(0.2405 × 540) + (0.00002143 × 356,400)] = 1.0064 ×
    137.51 = ...138 40
H$_2$O 0 0356 [(0 42 × 540) + (0 000185 × 356,400)] = 0.0356 × 292 73 = 10 42
Air 0 3015 [(0 2335 × 540) + (0 0000208 × 356,400)] = 0 3015 × 133 50 = 40 23

1.9044

Total available heat lost in products of combustion escaping at 600° = 267 90

From this it appears that
$$\frac{267.9 \times 100}{1147} = 23.4 \text{ per cent.}$$
of the heat available in the furnace gases. Assuming 10 per cent. additional to cover losses due to radiation, etc., the net efficiency of the stoves under these conditions cannot be far from 75 per cent., and the net effective heat per pound of gas is
$$\frac{1147 \times 75}{100} = \text{about 860 B. T. U.}$$

Gas Requirement.—To furnish the heat for the blast for one hour requires, $\frac{50,000,000}{860} = 58,150$ pounds of gas per hour. The total gas made per hour by a furnace making 400 tons of iron per day on 2240 pounds of fuel will be $571.2 \times 373\frac{1}{3} = 213,250$ pounds. The quantity of gas required by the stoves is evidently, $\frac{58,150 \times 100}{213,250} = 27$ per cent. of the total gas formed. On a fuel consumption of 2500 pounds, this requirement falls nearly to 25 per cent.; but with 2000 pounds it rises to 29 per cent. of the gas made. The consumption of 25 per cent. of the gas on a fuel consumption of 2240, on the other hand, would permit a blast temperature of only about 1100 degrees F., instead of 1200 degrees F., but at the same time the volume of brickwork could be reduced from 18,500 cubic feet to 15,200, which requires only about 260,000 9-inch bricks.

Ordinarily 1 square inch of gas burner-area is sufficient for 200 square feet of stove heating surface.

Stove Linings.—In lining the stoves the firebricks should be preceded by a layer of cement on the bottom to exclude moisture completely. Between the shell and the first row of bricks an expansion space of 2 to 2½ inches should be left from bottom to top. This space may be filled with slagwool or other loose material except at the bottom, where cement should be used. The temperature at the bottom is never high enough to cause much expansion, and the cement will prevent leakage if the lower plate rusts.

Stove bricks are not subjected to such severe temperature as hearth and bosh bricks, hence are not so refractory, nor should they be as hard as inwall bricks. They must be capable of with-standing changes in temperature without cracking, and of resist-ing the disintegrating action of gases, and must possess porosity sufficient to absorb and give out heat readily. Glazed or vitrified bricks do not take up heat rapidly, hence vitrification should be avoided. But it is desirable that the temperature of burning should be high enough to convert all of the iron oxides present in the clay into silicates, so that they may not be disintegrated by the gases. A tolerably refractory brick should be used for the combustion chamber, or it may become vitrified. For hot blast mains and gas flues a less refractory brick may be used A dense, hard brick is desirable whenever the gases carry much dust, such as in the downtake and gas flues.

Bustle Pipe.—The blast connections should be of such ca-pacity that there will be no throttling of the blast, or excessive loss of head due to friction. For this reason, the smallest area of the bustle pipe should never be less than the combined area of the tuyere openings, or 350 square inches, which is equivalent to 21 inches diameter in the clear. Allowance for a 9-inch lining would require a pipe at least 40 inches in diameter.

Blast Mains.—As regards the hot and cold blast mains, a suitable diameter may be approximated when the length and per-missable loss in friction have been determined.

If we assume, for example, a hot blast main, whose average length from bustle pipe to stove is 100 feet, and a cold blast main whose average length from stoves to engine is 150 feet and that the temperature of the air in each is 1200 degrees F. and 150 degrees F., respectively, at 15 pounds pressure, we may calculate approximately the area of pipe desirable to deliver 35,000 cubic feet piston displacement per minute.

Air at 60 degrees F. and atmosphere pressure weighs 0 076 pound per cubic foot. Air at 150 degrees F. and 15 pounds pressure weighs 0 130 pound per cubic foot. Air at 1,200 degrees F. and 15 pounds pressure weighs 0 048 pound per cubic foot. Whence it appears that the engine delivers

$$\frac{35,000}{60} = 583 \text{ cubic feet per second}$$

And the cold main transmits $583 \times \frac{0\,076}{0\,130} = 340$ cubic feet per second.

And the hot main transmits $583 \times \frac{0\,076}{0\,048} = 923$ cubic feet per second

If we assume further, for example, that the total drop in pressure between the engines and the bustle pipe must not exceed 1 pound, of which 3 ounces may take place in the stove, 3 ounces in the cold main, and 10 ounces in the hot main, and that all bends in the mains shall be less than 45 degrees, and hence negligible, except one in each, which shall be 90 degrees and equivalent in resistance to 25 feet additional pipe of about 2 feet diameter, then we have, according to the formula,

$$d = \frac{lv^2}{25,000\,p},$$

Where, d = diameter of pipe in inches,
l = length of pipe in feet,
v = number feet traveled by each gas particle per second.
p = loss of pressure in ounces per square inch,
for the cold blast main, if the velocity is 100 feet per second.

"Mechanical Draught," B. F. Sturtevant Co.

$$d = \frac{175 \times (100)^2}{25,000 \times 3} = \frac{1,750,000}{75,000} = 23.3 \text{ inches,}$$

or, virtually 2 feet, and for the hot blast main, if the velocity is 233 feet per second,

$$d = \frac{125 \times (233)^2}{25,000 \times 10} = \frac{6,786,100}{250,000} = 27.1 \text{ inches.}$$

or, practically, $2\frac{1}{4}$ feet.

Air Receivers.—Unless an air receiver equal in volume to at least four piston displacements is used to absorb the shock of the piston impulses, a cold blast main of three or four feet in diameter may be necessary to prevent the pipe whipping. A receiver is preferable, and the former prejudice against it is disappearing because liability to explosions is entirely nullified by proper arrangement of check-valves in the hot blast main.

BLOWING ENGINES.

In order to determine the blowing engine capacity required for the furnace, it is necessary to know the quantity of air and the blast pressure. The former is tolerably fixed for a given size of furnace, but the latter varies with furnace conditions. The average theoretical number of horse power needed under given conditions is represented approximately by the equation

$$\text{Horse-power} = \frac{\text{cubic feet per m.} \times \text{P. per square foot.}}{33,000}$$

if we assume that the pressure in the air cylinder is about equal to the minimum pressure at the tuyeres.

The following table, based on this formula, represents the approximate number of horsepower needed for various quantities of air at different minimum pressures:

Pressure per square Inch..	20,000	30,000	40,000	50,000	60,000	70,000
	Cubic feet of air per minute					
8 pounds...............	700	1,050	1,400	1,745	2,095	2,445
9 pounds...............	785	1,180	1,570	1,960	2,335	2,750
10 pounds...............	875	1,310	1,745	2,180	2,620	3,055
11 pounds...............	900	1,440	1,920	2,400	2,880	3,360
12 pounds	1,050	1,570	2,095	2,620	3,140	3,665
13 pounds...............	1,135	1,700	2,270	2,835	3,405	3,970
14 pounds...............	1,220	1,830	2,445	3,055	3,665	4,275
15 pounds...............	1,310	1,960	2,620	3,270	3,930	4,580
16 pounds...............	1,400	2,095	2,795	3,490	4,190	4,890
17 pounds............. .	1,485	2,225	2,965	3,710	4,450	5,195
18 pounds	1,570	2,355	3,140	3,930	4,710	5,500
Required number revolutions cylinder, 84 x 60 in.	54	81	108	135	162	189
Expected output per day, gross tons:						
2,000 lb. fuel per ton...	260	390	515	645	775	900
2,240 lb. fuel per ton...	230	345	460	575	690	800
2,500 lb. fuel per ton...	210	310	415	515	620	725

With a blowing cylinder 84 inches in diameter and 60 inches stroke and an allowance of 4 per cent. for clearance, each revolution of the engine will represent 370 cubic feet of piston displacement, and hence each 10,000 cubic feet will require about 27 revolutions of the engine.

In order to supply 35,000 cubic feet of air per minute from a cylinder 84 X 60 inches requiring 27 revolutions for each 10,000 cubic feet of air, a total number of 95 revolutions will be necessary. As 50 revolutions is about the limit of practicable air-valve speed, and hence for satisfactory filling of the air cylinder, it follows that two air cylinders at 47½ revolutions each will be necessary. This result may be attained by one cross-compound engine, with two air cylinders, or by two disconnected engines, a high and low pressure which may be run compound, with one air cylinder each. This allows for no spares, however, so an extra engine should be installed. At present, vertical cross-compound engines of the "steeple" type, having two air cylinders, are extensively used. They require the least floor space, but owing to their height are difficult to repair Moreover, the stopping of one engine puts two air cylinders out of use. There is at present a tendency to revert to the old long cross-head type of single engines, as being simpler and less wasteful when idle. A high and low pressure pair, with a spare high pressure duplicate is a convenient arrangement for single furnaces For a pair of furnaces one spare is sufficient. However, single engines with tandem air and steam cylinders are never smooth running, since there is no compensation for the throw of the cranks at each revolution. Smooth running may be promoted by the use of the *quarter-crank* principle, by which two cranks operating on the same shaft are set 90 degrees apart. This method may be applied to all types of engine, except the long cross-head. Horizontal or vertical-horizontal engines obviate excessive heights and consequent vibration, but they require considerable floor space.

Engines are now usually designed to run compound on 125 to 150 pounds initial steam pressure. They are governed to show reasonable economy at all blast pressures between 12 and 20 pounds per square inch, but to show maximum efficiency at 15 to 16 pounds. The usual minimum requirement for an air cylinder for blowing a coke furnace is 20,000 cubic feet air per minute, which is equivalent to about 53 revolutions with an air cylinder 84 inches in diameter and 60-inch stroke. With most air valve gearing 53 revolutions is above the limit of speed for effective action. In consequence air cylinders less than 84 inches in diameter have become practically obsolete, and some makers advocate 96-inch cylinders with slower action. .

As steam pressures at a furnace plant are dependent largely upon the condition of the furnace action, and consequently upon the supply of gas, it naturally follows that they frequently fall below 125 pounds. Many engines are therefore designed to run on steam pressure as low as 100 pounds In order that the mean effective pressure in the low pressure steam cylinder, run with a good vacuum, may always overcome the resistance offered by the back pressure of the blast, even when the steam pressure is as low as 100 pounds, the ratio of the two steam cylinders should be small, as, for example: 1 : 3.6 or 1 : 3.3. For a low pressure cylinder, 84 inches diameter, this corresponds to high pressure cylinders of 44 or 46 inches diameter. With steam at 150 pounds, a ratio of 1 : 4 is perfectly efficient and permits the use of a 42-inch cylinder, but the engine may fail on low steam pressures. It is better, therefore, to use large high pressure cylinders in order to be prepared for low steam pressure and to control the efficiency

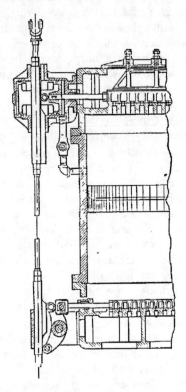

The Southwark Air Cylinder, Showing Inlet and Outlet Valves, with Gear.

of the engine when running on high steam pressures by varying the high pressure cut-off in such a way as to give the low pressure cylinder steam at a constant pressure, regardless of the initial steam pressure. In a low pressure cylinder of the same area as the air cylinder, the initial pressure should not fall much below 50 pounds.

The chief distinguishing characteristics of the different makes of blowing engines lie in the air end, and are centered in the design and action of the air valves. In order that air cylinders may furnish blast on both strokes of the piston they must be fitted with an inlet and an outlet valve at each end. These are usually constructed in the cylinder heads and approximately half the head is devoted to each valve.

The air valves of the **Southwark** engines are of the rectangular gridiron type and are opened and closed by a small lateral travel of the valves. The inlet valve is operated positively throughout by means of a straight cam shaft. It starts to open on the dead centre, and is fully open at 10 per cent. of the stroke. It begins to close at 90 per cent. of the stroke, and is therefore open wide and the valve is stationary for 80 per cent. of the stroke. As both opening and closing are absolutely positive, the action is equally effective at all speeds.

The outlet valve is closed positively by means of a straight cam, but is opened automatically by the action of the blast pressure. This is accomplished by means of a by-pass pipe, leading from the interior of the air cylinder to a small auxiliary cylinder, whose piston is fitted to an extension of the valve stem. The outlet valve is slightly smaller than the inlet.

These valves are guaranteed to operate efficiently at 80 revolutions per minute, and are, therefore, well adapted for use in gas blowing engines which work best at high speeds. They are used in connection with the Koerting engines which operate the furnaces of the Lackawanna Steel Company at Buffalo.

The air cylinder of the **Mesta** Machine Company's blowing engine has a positive acting rotary inlet valve of the Corliss type, extending across the cylinder head, and operated by means of a wrist-plate. The outlet valves are circular poppets, usually three in number. They are closed mechanically, but opened automatically by the blast pressure when the pressure in the cylinder equals that in the blast main. The valve stem carries a small piston, working in a dash-pot, which cushions the opening movement. The closing is accomplished through the operation of the wrist-plate. When the stroke is nearly complete, a sleeve containing a spiral spring engages a collar on the valve stem and forces the

valve to its seat just as the piston reaches the end of the stroke.
The area of the inlet openings is about 12 per cent., and that of
the outlet about 10 per cent. of the piston area. The maximum
speed of operation claimed is 60 revolutions per minute.

The valves used on the air cylinders of the **Tod** engines are
circular, there being usually two inlet and three outlet valves in
each cylinder head. The inlet valves consist of double ported
pistons working in cages set in the cylinder heads. The valve is
operated positively by means of levers attached to a wrist-plate

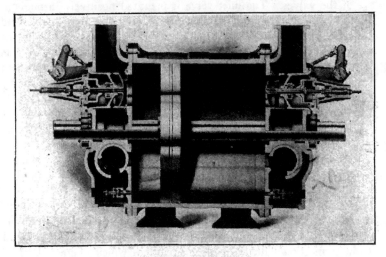

The Mesta Air Cylinder.

which opens it when the pressure falls to that of the atmosphere
and closes it at the dead centre. By means of an adjustable link
the time of opening is varied to correspond with the pressure of
the blast.

The outlet valve is of the poppet type which opens automatically when the pressure in the cylinder equals that in the blast
Iron Age,
Nov. 2, 1905,
p. 1149. mains. It is closed positively by means of a lever operated by the
wrist-plate. The area of the valves may equal 12 per cent. of the
piston area and they can be operated at 50 revolutions.

The **Kennedy-Reynolds** vales are used on the Allis-Chalmers
·engines. The Kennedy valve is the inlet, and consists of a hollow

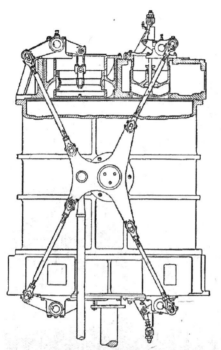

Tod Air Cylinder, Showing Operation of Valves.

Head of Air Cylinder and Parts of Inlet and Outlet Valves of Tod Blowing Engine.

Age, 1906, 1053.

cast iron tube passing through the centre of the cylinder. It is somewhat objectionable, as it allows leakage, owing to the friction of rubbing, and also necessitates the use of two piston rods. The outlet or Reynolds valve is cup-shaped, free to open automatically, but is closed positively. Their area is fully equal to 8 per cent. of the piston area, and they can be operated safely at 30 revolutions per minute.

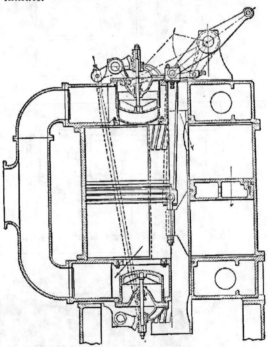

The Kennedy-Reynolds Valve.

The **Weimer** blowing cylinder is provided at each end with a peripheral extension ring, having an A-shaped cross-section. The ring is pierced on each slope by a series of slots ¾ inch wide, closed by aluminum strips. The slots on the outer slope act as inlet valves and those on the inner slope as outlet valves. Each set of valves equals 19 per cent. of the cylinder area, and they are all operated automatically by the pressure of the air.

POWER REQUIREMENT.

Blowing.—In the case of a furnace, for example, which requires 35,000 cubic feet of air per minute at a minimum pressure of 15 pounds per square inch at the furnace, we see by the table that the air requirement will demand about 2300 H. P. theoretically at the furnace, or about 2500 at the engine. One theoretical H. P. per minute requires 42.42 B. T. U., hence a total of 42.42 × 2500 = 106,000 B. T. U. per minute, or 6,360,000 B. T. U. per hour will be required to furnish the blast.

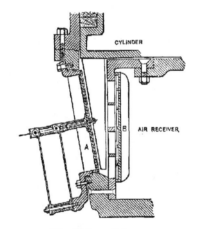

The Welmer Valve.

Owing to lack of economical efficiency in engines and boilers, the actual requirement of power is much greater. It may be estimated approximately as follows:

Assuming a compound condensing engine, when running on steam at 150 pounds pressure, to consume about 16 pounds of steam per indicated H. P., utilizing 18.75 B. T. U. per pound of steam per minute. 16 × 18.75 =300 B. T. U. per I. H. P. per minute or 18,000 B. T. U. per hour. As 42.42 × 60 = 2545 B. T. U per hour is a theoretical H. P., this indicates a thermal efficiency of

$$\frac{2545 \times 100}{18,000} = 14.14 \text{ per cent.}$$

Reckoning a mechanical efficiency on both ends of the engine of 87 per cent of the indicated thermal efficiency gives an actual thermal efficiency of 12.3 per cent. Assuming a boiler efficiency of 65 per cent, the fuel efficiency of the system would be 12.3 × 65 = 8 per cent. The quantity of heat required, therefore, to blow the furnace per hour will be

$$\frac{6,360,000 \times 100}{8} = 79,500,000 \text{ B. T. U.}$$

$$\frac{79,500,000}{1147} = 69,300 \text{ pounds of gas.}$$

Hoisting.—The power expended in hoisting the stock may be found as follows. Under the present assumption the quantity of stock raised per ton of pig is,

Coke	1.0 tons.
Ore	1.7 tons.
Stone	0.5 tons
Total	3.2 tons

The quantity of stock required each hour by a 400 ton furnace will be $3.2 \times 2240 \times \frac{400}{24} = 119,400$ pounds. Assuming for an 80 foot furnace that the total vertical height from the stockhouse floor to the dumping point is 110 feet, then, neglecting the weight of the skips since they should be balanced, the total work required will be, $119,400 \times 110 = 13,134,000$ foot pounds per hour. $\frac{13,134,000}{778} = 16,880$ B. T. U. Assuming the efficiency of the hoisting engine to be 10 per cent of that of the blowing engine, this would require an expenditure of $\frac{16,880 \times 100}{0.8} = 2,110,000$ B. T. U., which is equivalent to $\frac{2,110,000}{1147} = 1840$ pounds of gas per hour.

Pumping.—Assuming that 10,000 gallons of cooling water are required per ton of iron produced, then 4,000,000 gallons are needed by the furnace in 24 hours, and the power required to pump this quantity may be determined as follows:

$$\frac{4,000,000 \times 8\frac{1}{3}}{24} = 1,400,000 \text{ pounds of water per hour.}$$

A 25 pound pressure at the tuyeres requires a head of 65 feet above the working level of the furnace. Assuming the source of the water supply to be 35 feet below this level, there will be a total lift of 100 feet. The total work done by the pumps per hour will then be $1,400,000 \times 100 = 140,000,000$ foot pounds

$$\frac{140,000,000}{778} = 180,000 \text{ B. T. U.}$$

Assuming a pump efficiency of 20 per cent. which is .23 of that of the blowing engine, we will have a thermal requirement of

$$\frac{180,000 \times 100}{0.23 \times 8} = 9,782,600 \text{ B. T. U., which equals,}$$

$$\frac{9,782,600}{1147} = 8530 \text{ pounds of gas per hour.}$$

Feed Water.—If the blowing engine uses 16 pounds of steam per H. P. per hour, the quantity of feed water required will be

$$16 \times 2500 = 40,000 \text{ pounds per hour,}$$

but steam required for other uses will bring this amount up to 50,000 pounds at least. The total head against the pumps will be something in excess of the boiler pressure of 150 pounds. Let us assume 160 pounds for example. Then the total work done by the pumps will be,

$$50,000 \times 160 = 8,000,000 \text{ foot pounds per hour.}$$

$$\frac{8,000,000}{778} = 10,280 \text{ B. T. U.}$$

If the efficiency of the pumps is 10 per cent., this work will require

$$\frac{10,280 \times 100}{0.8} = 1,285,000 \text{ B. T. U.}$$

which equals $\dfrac{1,285,000}{1147} = 1120$ pounds of gas per hour.

Lighting.—The power necessary to light a plant by means of electricity and to run an ore bridge, larry and machine shop may be estimated if the number and character of lights and the power of the motors is known. Assuming that 25 arc and 100 incandescent lights are used, then

$$25 \times 700 = 17,500 \text{ watts}$$
$$100 \times 50 = 5,000 \text{ watts}$$
$$\text{Total power} = 22,500 \text{ watts.}$$
$$\text{One watt hour} = \frac{2,545}{746} = 3.4 \text{ B. T. U.}$$
$$3.4 \times 22,500 = 76,500 \text{ B. T. U.}$$

Assuming the same efficiency as the blowing engine, gives

$$\frac{76,500 \times 100}{8} = 956,250 \text{ B. T. U. per hour}$$

$$\frac{956,250}{1,147} = 835 \text{ pounds gas per hour}$$

Power.—Assuming 5-50 H. P. motors for the other uses, we have,

250 × 2,545 = 636,250 B. T. U., which at the same efficiency will require

$$\frac{636,250 \times 100}{8} = 7,953,150 \text{ units, or}$$

$$\frac{7,953,150}{1,147} = 6,935 \text{ pounds of gas}$$

Summary.—The total power developed and gas consumed may be summarized thus:

	H. P	Pounds gas per hour.
Blowing	2,500	69,300
Hoisting	70	1,840
Cooling water	310	8,530
Feed water	40	1,120
Lighting	30	835
Power	250	6,935
Totals	3,200	88,560

which is about 8 H. P. per ton of coke burned. Gas required per

$$\text{H. P.} = \frac{88,560}{3200} = 27.7 \text{ pounds.}$$

Gas Distribution.—The percentage of the total gas generated with a fuel consumption of 2240 pounds coke per ton of pig which is consumed in power development, is

Per cent

$$\frac{88,560 \times 100}{213,250} = \dots \dots \dots 41\ 5$$

Percentage of gas utilized in heating blast = ...27.0

Total usefully applied = ... 68 5

Leaving for other purposes and losses = ...31.5

The following summary is designed to show the thermal efficiency of the blast furnace under various fuel consumptions:

Fuel consumption	2,000 lb	2,240 lb	2,500 lb
Carbon burned to CO per 100 pounds pig	39 lb	46 9 lb	56 lb
Carbon burned to CO_2 per 100 pounds pig	29 lb	29 0 lb	29 lb.
Total carbon burned per 100 pounds pig	68 lb	75.9 lb.	85 lb.

Total heat developed per 100 pounds pig	595,500 B. T. U.	630,655 B. T. U.	671,150 B T. U.
Total heat possible per 100 pounds pig	989,400 B T. U	1,104,345 B T. U.	1,236,750 B T U.
Under ordinary conditions			
Heat utilized in the furnace	591,000 B T. U.	591,000 B T. U.	591,000 B. T. U.
Heat utilized in the stoves	164,920 B. T. U	177,300 B. T. U.	189,000 B. T. U.
Heat utilized in the boilers	250,230 B. T. U.	272,500 B T. U	294,840 B. T. U.
Total heat usefully applied	1,006,150 B. T. U.	1,040,800 B. T. U.	1,074,840 B. T. U.
Percentage of possible heat	100.2	94.2	86.9
Under ideal conditions:			
Heat utilized in the furnace	591,000 B. T. U.	591,000 B. T. U.	591,000 B. T. U.
Heat required for the blast	108,150 B. T. U	120,720 B. T. U.	135,200 B. T. U.
Heat required for power	73,300 B. T. U	81,440 B T. U.	91,210 B. T. U.
Total heat required	772,450 B. T. U.	793,160 B. T U.	817,410 B. T. U.
Percentage of possible heat	78.1	71.8	66.1

BOILERS.

The boiler capacity necessary to run all the forms of power development of a 400 ton furnace is, as we have seen, 3200 H. P. The type of boiler best suited to gaseous fuel is the water tube, and its use has become practically universal for blast furnace work. The make most generally used is probably the **Babcock and Wilcox,** although the Stirling is becoming very popular also. They are both of the horizontal drum variety. The **Stirling,** however, has curved tubes which do not lend themselves readily to cleaning. The **Cahall** and **Wheeler** makes are watertube vertical boilers, and are also considerably used. The Cahall is expensive to maintain and is not growing in favor. The Wheeler is cheaper to install, gives less trouble, and stands more abuse.

The boiler plant should always have at least one spare unit for use during cleaning and repairs. The size of unit should range from 200 to 400 H. P., according to the size of plant. The unit should not be so large in proportion that one more or less seriously affects the steam capacity. Experience shows that never less than 6 to 8 H. P. is needed per ton of iron. When pressure is above 12 pounds, and electric machinery has to be operated in the yard, 8 to 10 H. P. are needed. Ordinarily 1 square inch of gas burner area is sufficient to maintain 1 boiler H. P.

PUMPS.

The piston pump has long been used to furnish **cooling water** to the blast furnace plant, but of late there is a growing application of the centrifugal pump. Of whatever form is used, there should not be less than three units, of which two are of the requisite capacity and the third is a spare. To furnish 4,000,000 gallons daily, three two million gallon pumps should be provided. A 14 x 10 plunger operated at 54 revolutions will furnish 2,000,000 gallons in 24 hours.

For **boiler feed,** the positive acting piston pump is required. The plunger pump with outside packing is most desirable. Two units are sufficient under ordinary conditions. Two plungers, 6 x 10 inches at 35 strokes per minute, will furnish feed water for a 400 ton furnace.

GAS ENGINES.

Owing to the admittedly low thermal efficiency of steam equipment in general, and to the fact that blast furnace gas is well adapted to direct combustion in gas engines, there has been considerable application of gas engines to blast blowing work.

It is found as a rule that about 11,500 B. T. U. suffices to develop a brake horse power when its energy is expended in the cylinder of a gas engine. This indicates a thermal efficiency of

$$\frac{2545 \times 100}{11,500} = 22 \text{ per cent.}$$

$$\frac{11,500}{1057} = 10.9 \text{ pounds of gas per H. P., which is } \frac{10.9 \times 100}{27.7}$$

$= 39$ per cent. of that required by the steam engine per H. P.

$$10.9 \times 3200 = 34,880 \text{ pounds gas per hour, or } \frac{34,880 \times 100}{213,250}$$

$= 16.35$ per cent. of the gas generated.

41.5 per cent. — 16.35 per cent = 25.15 per cent, or over ¼ of the total gas may thereby be saved for other uses. 25.15 per cent. + 31.5 per cent. makes a total of 56 65 per cent. of gas which might be utilized for other purposes. The quantity of gas so wasted equals,

$$213,250 \times .5665 = 120,800 \text{ pounds.}$$

If 10.9 pounds gas, exploded in a gas engine, is capable of producing 1 H. P., then the power so saved will equal,

$$\frac{120,800}{10.9} = 11,080 \text{ H. P., or about 28 H. P. per hour}$$

TABLE SHOWING THE COMPARATIVE EFFICIENCY OF GASES RESULTING FROM VARIOUS FUEL CONSUMPTIONS IN A 400-TON FURNACE

	Fuel consumption per ton of pig made		
	2,000 pounds	2,240 pounds	2,500 pounds
Quantity of gas generated per second	33.68 lb	59.23 lb	65.34 lb
Quantity of gas generated per pound pig	703.0 cu. ft.	776.0 cu. ft.	856.0 cu. ft.
Sensible heat per pound of gases at 450 degrees F.	5.17 lb / 67.7 cu. ft.	5.11 lb / 74.8 cu. ft.	6.30 lb / 82.5 cu. ft.
Potential heat per pound of gases still undeveloped	100 B.T.U.	109 B.T.U.	100 B.T.U.
Potential heat per cubic foot of gases still undeveloped	1,000 B.T.U.	1,050 B.T.U.	1,100 B.T.U.
•	75 B.T.U.	80 B.T.U.	85 B.T.U
Proportion of gases needed to heat blast to 1,200 degrees F.	29 per cent	27.0 per cent	25 per cent
Proportion of gases needed to raise steam	44 per cent	41.5 per cent	39 per cent
Proportion of gases unused by plant	27 per cent	31.5 per cent	36 per cent
•	100 per cent.	100.0 per cent.	100 per cent.
Proportion of gases needed for steam power.	44 per cent.	41.5 per cent	39 per cent
Proportion of gases needed by gas engines.	19 per cent	16.5 per cent	14 per cent
Net saving by substituting gas engines.	25 per cent	25.0 per cent	25 per cent
Gas required per H.P. with steam engines	29.30 lb / 384.0 cu. ft.	27.70 lb / 363.0 cu. ft.	26.20 lb / 343.0 cu. ft.
Gas required per H.P. with gas engines	11.55 lb / 151.0 cu. ft.	10.90 lb / 143.0 cu. ft.	10.25 lb / 134.0 cu. ft.
Net thermal efficiency of steam engines	8.0 per cent	8.0 per cent	8.0 per cent
Net thermal efficiency of gas engines.	20.0 per cent	20.0 per cent	20.0 per cent
H.P. per ton pig saved by substituting gas engines	12.5	12.5	12.5
H.P. per ton pig previously unused	12.0	15.5	19.0
Total H.P. per ton pig available for other uses	24.5	28.0	31.5

per ton of pig made. However, the saving which can be credited purely to the use of gas engines for necessary power development will be only

$$\frac{25.15}{56.65} \times 28 = 12.5 \text{ H. P per ton of product,}$$

since the remainder was already in excess of the requirement.

The **Koerting** Gas Engine, built by the De La Vergne Machine Company, is a two-cycle, double-acting engine, taking impulses on both strokes in the same manner as a steam engine. Gas and air are forced by separate pumps into each end of the cylinder alternately. They mingle as they enter and are ignited by electric sparks. The exhaust openings are in the periphery of the cylinder at the middle, so that, owing to the length of the piston, they are not uncovered until toward the end of the stroke, thereby obviating the necessity for exhaust valve mechanism.

In the diagram the power piston is represented as just starting toward the rear end of the cylinder and the pump pistons are just starting toward the crank end. Explosion has just taken place. Gas and air are being forced into the connecting passages under compression. The admission valve, however, is held closed by the high pressure in the cylinder and further ingress is impossible

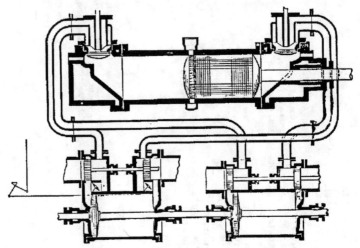

Diagram Showing Arrangement of Gas and Air Ducts of Koerting Double Acting Gas Engine.

until the exhaust ports are uncovered. The fresh air and gas then rush in, and sweep out the products of the previous explosion. They are compressed again by the returning piston, preparatory to being ignited in turn.

The two-cycle engine shows an average mechanical efficiency of only 75 to 80 per cent owing to the expenditure of power in operating the gas and air pumps. The slightly higher efficiency of four-cycle engines is offset, however, by the liability to derangement of their highly complex valve mechanism. On the other hand the total number of cylinders involved in the four-cycle engine is less, and there is at present a tendency to adopt it for blast furnace work.

With blast furnace gas giving a mean effective pressure of about 65 pounds per square inch, 1000 Brake H. P. will be delivered by a double-acting cylinder 38 x 60 inches operated at 75 revolutions. Such a cylinder could drive a 72 inch air cylinder delivering 20,000 cubic feet of air against 15 pounds pressure, or a 60 inch cylinder delivering 14,000 cubic feet at 25 pounds pressure.

In the face of emergencies, however, gas engines lack the flexibility of steam engines. They have little or no overload capacity. Excessive demands on a steam engine may be met by higher steam pressures or later cut-off. With the gas engines the quantity of gas and pressure of explosion is tolerably constant. It is necessary, therefore, to design a gas engine for the maximum duty required of it, and operate it as near full load as possible. Hence, an 84-inch air cylinder at 15 pounds pressure can be operated by a 42-inch gas cylinder, whereas a pressure of 25 pounds demands a cylinder 54 inches in diameter, which must be provided if such a contingency is to be met by a single engine. Since the maximum economy of the gas engine is realized only when it is operated at full load, it is better to design engines for ordinary loads and to meet emergencies by utilizing reserve units.

GAS WASHING.

Since gas engines require a gas which is practically free from solid matter an efficient system of gas washing is necessary. The quantity of dust remaining in the gases should not exceed 0.1 grains per cubic foot.

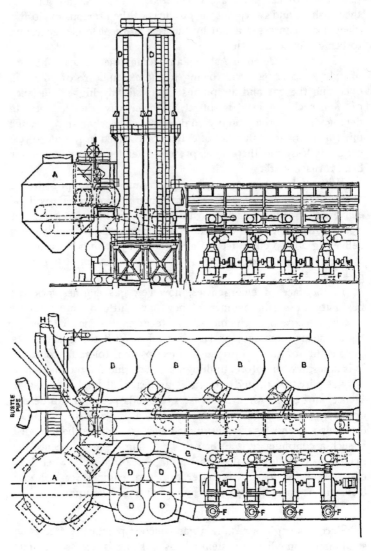

Gas Cleaning Plant at the Lackawanna Steel Company.

The gas-washing apparatus used at the **Lackawanna** plant in Buffalo consists essentially of a dust catcher (A), followed by cooling towers (D), which contains Koerting water-spray nozzles, and by hydraulic fans (F) in which the gas is beaten violently with water. The gas is then led away through pipes (G. H.) to engines and stoves (B). The cleaned gas is said to contain only 0.02 grains solid matter per cubic foot of gases.

The **Bian** apparatus used in Germany reduces the dust to 0 01 grains per cubic foot of gases. It consists essentially of perforated discs, rotating 10-12 times per minute in a tank half filled with water, through whose moistened perforations the gas is forced by means of a fan, followed by filtering towers to remove the moisture. The cost of installation is about $60 per 1000 cubic feet of gas cleaned, the water consumption is 12 gallons, the H P. required 0.3, and the cost of operating 7¼ cents.

The **Theisen** washer, which is in high favor abroad, consists of a cylinder supplied with water, into which the gas is forced while the cylinder is rotated 850 times per minute. The purification is very complete at moderate cost. Per 1000 cubic feet of gas the costs are said to be as follows: Installation, $23; water consumption, 9 gallons; power consumed, 0.15 H. P.; operating expenses, 3 8 cents.

ROLLING STOCK.

All rolling stock equipment should be standard gauge, and fitted with standard and interchangeable parts, such as trucks, couplers, etc.

Hot metal and cinder cars should have heavy framework, firmly bolted together, connecting the trucks. Structural weakness may have disastrous results.

Iron Ladles.—Hot metal requires 5 cubic feet space per ton, and a 20-ton car, therefore, must have 100 cubic feet net capacity. A car having 5 feet average diameter and 5 feet available depth will hold 20 tons. About 20 per cent. extra depth should be allowed, however, for skulling and to prevent slopping on curves and grades. The lining must be of refractory bricks and not less than 4½ inches thick. For tapping a 400-ton furnace every six hours, five such ladles will be required, not counting spares.

Cinder Ladles.—The molten cinder requires about 15 cubic feet per ton and a 15-ton cinder car requires 225 cubic feet net capacity. A bowl having 7 feet average diameter and 6 feet available depth will contain 230 cubic feet. The cinder bowl may be lined with firebricks, or with a cast iron thimble which is removable. The latter, however, is of no avail when molten iron is tapped with the cinder, as it cuts through the thimble and the cinder flows over the tracks. In dumping, the ladles should always travel forward, in order to clear the track. If the furnace makes half a ton of cinder for each ton of iron, or 50 tons in six hours, it will evidently require the pretty constant services of at least four cinder cars.

SUPPLEMENT

USES OF PIG IRON.

GRADES OF PIG IRON.

In the manufacture of ferrous products pig iron exhibits a wide range of usefulness. It is not only used practically unchanged to produce castings of a great variety of form and quality, but it is the starting point of all present methods of producing wrought iron and steel. Since each subsequent operation demands a composition which is within certain limits, it follows that pig iron must necessarily offer a considerable variety of composition. The usually specified limits of the chief grades of pig iron are as follows:

Grade of Iron.	Si, per cent.	S, per cent.	P, per cent.	Mn, per cent.
No. 1 foundry	2.50 to 3.00	under 0.035	0.5 to 1.0	under 1.0
No. 2 foundry	2.00 to 2.50	under 0.045	0.5 to 1.0	under 1.0
No. 3 foundry	1.50 to 2.00	under 0.055	0.5 to 1.0	under 1.0
Malleable	0.75 to 1.50	under 0.050	under 0.2	under 1.0
Gray forge	under 1.50	under 0.100	under 1.0	under 1.0
Bessemer	1.00 to 2.00	under 0.050	under 0.1	under 1.0
Low phosphorus	under 2.00	under 0.030	under 0.03	under 1.0
Basic	under 1.00	under 0.050	under 1.0	under 1.0
Thomas Gilchrist or basic Bessemer	under 1.00	under 0.050	2.0 to 3.0	1.0 to 2.0

CLASSIFICATION OF PIG IRONS.

According to their uses, pig irons may be separated roughly into two great classes, first, those which undergo complete conversion into other forms of ferrous products, and, second, those which are not materially changed in composition or nature.

In the first class are included all grades which are intended primarily for use in the manufacture of wrought iron and steel, such as gray forge, Bessemer, basic, low phosphorus and Thomas irons. In the second class are the foundry and malleable grades, although, as we shall see later, the malleable grade properly stands between the foundry and conversion irons in this classification.

CONVERSION OF PIG IRON.

The object of conversion is identical in all cases. It consists essentially of the elimination of as much as possible of all the non-ferrous elements which are present in the pig. Differences in the resulting products naturally follow from differences in the converting processes employed. The most obvious distinction between the various converting processes is based upon the state of the product. When the resulting metal is in a completely fused state it receives the name of steel, when in a pasty or semi-fused state it is known as wrought iron. This difference of condition in conversion gives to these two types of product physical characteristics which differ far more than the chemical compositions, and is therefore the basis of distinction between them in this country.

WROUGHT IRON CONVERSION.

The process of conversion which has wrought iron for its product is known as **puddling.** The grade of pig which is designed primarily for this process is gray forge. The operation is carried on in a **reverberatory furnace,** which consists essentially of a low rectangular chamber built of firebrick and having a low dividing wall, cutting off about one-third of its length. The larger portion is the working chamber and the smaller is the fireplace. The fuel used is long flame bituminous coal. The flame passes over the dividing wall or **bridge** and heats the working chamber by radiation on its way to the chimney. In this way the charge is kept from contact with the fuel itself

Nature of Hearth.—The composition and arrangement of the hearth material of the puddling furnace are of first importance to the process. The hearth bottom is usually made up of some refractory material, over which a layer of roll-scale is spread and made to soften and cohere by firing. Then a ball of wrought iron scrap is worked back and forth over the hearth at welding temperature, thereby giving it a smooth, even coat of magnetic oxide. The side walls are then **fettled.** The fettling consists of oxides of iron, usually in the form of dense, lumpy ore, either hematite or magnetite. It is arranged around the walls of the furnace at the slag level in such a way as to give the hearth a dished shape. The fettling serves a double purpose: it protects the firebrick walls from the corrosion of the slag, and it furnishes oxygen for oxidizing the metalloids of the charge. This is probably acomplished chiefly through the medium of the slag, which acts as a carrier of the oxygen in the form probably of Fe_3O_4.

Operation. —The charge for two men consists of about $\frac{1}{2}$ ton of pig iron. Its introduction into the furnace is preceded by 50 to 100 pounds of slag from a previous heat, in order to insure slag early in the operation. The melting stage occupies about half an hour, at the end of which the metal is in a state of quiet fusion. During this stage the greater part of the silicon and manganese and some of the iron have been oxidized to SiO_2, MnO, and FeO and Fe_2O_3, respectively, and have united with each other to form a slag which is a silicate of iron and manganese. By means of a rabble, the whole charge is then stirred, to bring the metal into contact with the fettling, the newly formed slag and the air, by which the carbon can be oxidized. CO is formed and given off as bubbles, which give the bath the appearance of boiling. The charge foams up and some of the slag overflows. Then the reaction becomes less violent and gradually subsides as the carbon is eliminated and the iron is said to "come to nature." This iron is no longer fusible, but exists as globules of metallic iron whose melting point is above the temperature of the furnace. It is necessary then to collect these detached masses into balls. The time of heat is $\frac{1}{2}$ to 2 hours. The balls are withdrawn from the furnace dripping with slag, which fills all of the interstitial spaces, and are sent to a hammer or "squeezer." Here the bulk of the slag is squeezed out and the ball compacted into a kind of billet which goes at once to the rolls and is rolled into a flat bar, about $3\frac{1}{2}$ x $\frac{5}{8}$ inch in section, which is known as **"muckbar."** The muck bar is subsequently sheared, piled and rerolled into **finished iron.**

Elimination of Metalloids.—In the order of elimination, **silicon** stands first. The elimination is accomplished through the oxidation of Si to SiO_2, in which condition it can no longer unite with the metal, but must transfer itself to the slag. This operation is generally complete by the time fusion is accomplished. The oxidation of silicon is more rapid and complete at low temperatures, because at high temperatures oxygen shows a preference for carbon. At low temperatures carbon will not be attacked as long as silicon is present.

It is apparent that the presence of much silicon in forge irons
has several disadvantages. In the first place, as just stated, it
defers the elimination of carbon and prolongs the operation by
just that amount. In the second place, it makes more slag The
quantity of slag is determined by the quantity of silicon present
The slag is mainly ferroso-ferric silicate, having about 30 per cent
SiO_2 and 50 per cent metallic iron The ratio of Si to Fe is there-
for about 1·4. Consequently, every pound of Si in the slag means
also the presence of 4 pounds of Fe This large apparent loss is
partly compensated, however, by the fact that the oxidation of Si
by Fe_3O_4 of the slag according to the reaction,

$$Fe_3O_4 + 2Si = 3Fe + 2SiO_2,$$

yields 3 pounds Fe for every pound Si oxidized. A third disadvantage of much Si in the pig is due to the fact that highly siliceous slags are less able to carry P and S than basic slags. To be
sure, a certain quantity of slag is essential to the well-working
of the process, but it is very evident that any excess is not simply
a loss of so much metal to the purchaser, but is positively detrimental.

The elimination of **manganese** follows that of silicon so closely
that it is almost simultaneous with it. Like silicon, it is practically all eliminated and must therefore be reckoned as a dead
loss in purchasing. Its elimination is accomplished by oxidation
also. While all of the Mn is oxidized, its oxidation is only partial since most of it enters the slag as MnO, instead of MnO_2.
MnO acts as a base and unites with SiO_2. It does not tend to
increase the quantity of slag as SiO_2 does, but simply helps to
satisfy the SiO_2, thereby liberating an equivalent weight of iron
from the slag When the manganese is oxidized by the oxides in
the slag, according to the reaction,

$$Fe_3O_4 + 4Mn = 3Fe + 4MnO,$$

each pound of Mn yields ¾ pound of Fe. The gross saving for
each pound of Mn, therefore, may amount to 1¾ pounds Fe.

The elimination of **phosphorus** is not so rapid as that of silicon
and manganese, nevertheless about half is removed during the
melting period It must be eliminated fairly early in the operation, while the temperature is low. The removal of phosphorus

is also effected through oxidation, by which it enters the slag, but at high temperatures oxygen shows a preference for carbon, and it will desert the phosphorus which has been oxidized, leaving it in the elemental condition, so that, perforce, it must return to the metal. Its elimination is facilitated by removing before the temperature rises too high the slag first formed, which contains the greater part of silicon and phosphorus. The slag subsequently formed will necessarily be more basic on account of the absence of silica, and will therefore hold the oxides of phosphorus more tenaciously. Even under the most favorable conditions, the removal of phosphorus is never as complete as that of silicon or manganese. Probably never more than 90 per cent., and generally not over 75 per cent, of the phosphorus is removed in puddling. It is probable, however, that much of the remainder exists in the slag in the oxidized condition and so does not exert its usual deleterious effects on the metal. The oxidation of phosphorus by the slag, according to the reaction,

$$5Fe_3O_4 + 4P_2 = 15Fe + 4P_2O_5,$$

shows a yield of over 3 pounds of iron for every pound of phosphorus oxidized.

The elimination of **sulphur** is the last and the least satisfactory of all, for which reason it is highly desirable that it should be kept low in the blast furnace, which is the best place for its elimination. In the puddle furnace the average elimination is probably not over half of the quantity present. Sulphur may be eliminated either in the elemental or oxidized condition—either by volatilization or by scorification. The elimination is very slight at first, but increases rapidly toward the end. This is probably due to the fact that sulphur is decidedly volatile and cannot resist the higher temperatures of the later stages. It is also carried away by a basic slag which we have seen to be favored toward the end of the operation. The presence of manganese favors the removal of sulphur, owing to the stability of manganese sulphide.

Influence of Slag.—We have seen that an excessive amount of slag is likely to be a cause of serious waste in puddling. When properly proportioned, however, it is capable of yielding a compensatory return of metal through the reduction by the metalloids of the oxides of iron derived from the fettling. For this reason it is usual for the process to make on the average a yield of puddled bar nearly equal to the weight of pig used. The proper amount of slag is the least quantity that will carry the burden of metalloids and properly protect the iron from oxidation. An attempt to puddle too dry with a deficiency of cinder will leave the iron exposed to the oxidizing flame. As a result, much will be lost through the stack as fumes of iron oxide. On the other hand, the oxidation of the remaining metal will not be properly neutralized by the scanty supply of slag, and even if it can be properly balled up, the resulting bar will probably show lack of cohesion, being **red short** or **cold short** or both.

Character of Product.—Bars of wrought iron which have been produced by rolling puddled balls show certain characteristics which distinguish them from their counterpart, mild steel. A freshly sheared or fractured surface will show small black spots, which are never present in steel. These spots are the exposed ends of filaments of slag which have been drawn out by rolling from the globules imprisoned in the squeezed ball. The fineness of the markings serves as a rough guide to the amount of rolling to which the piece has been subjected. Again, if a bar be nicked and bent cold, the fracture at the bend will show a silky lamination or fibrous structure, while the steel fracture is always crystalline. The fibrous structure also is attributed to the drawing out of globules of metal, interspersed with particles of slag. The quantity of slag so imprisoned is never as great by weight as it appears, since the slag is of much lower specific gravity than the iron. The quantity may range from 0 2 per cent. to over 2 per cent. by weight, depending upon the care taken in making and refining the iron. A considerable proportion of the phosphorus and sulphur which analysis shows to be in wrought iron is present in the oxidized condition in this slag. It is, therefore, not injurious, as when combined with the metal. It was formerly argued that the superior welding qualities of wrought iron are due to the presence of this slag, which acts as a flux for any scale or oxide films that may cover the surfaces during heating and interfere with perfect metallic contact. Experiment shows, however, that high slag irons do not weld any better than those containing little slag. It is difficult to see how such action could well be expected of a silicate which is already saturated with oxides of iron.

Composition of Product.—The analysis of wrought iron may, as a rule, be expected to fall within the following limits:

C.	Sl.	Mn.	P.	S.	Fe.
0.05-0.25	0.02-0 2	Tr.-0.1	0 05-0.2	0 02-0 1	99.8-99.0

From the preceding discussion, therefore, it is evident that the initial amount of P should not exceed 1 per cent.; that S should not rise much above 0.1 per cent.; that silicon should be limited to what is needed to create sufficient cinder, and that manganese might as well be absent.

STEEL CONVERSION.

There are four chief grades of iron which have conversion into steel as their ultimate design—namely, Bessemer, low phosphorus, basic and Thomas-Gilchrist grades. They are each designed for a certain method of conversion and are not adapted, generally, to use in any of the other methods. The **Bessemer** grade is naturally designed for the Bessemer process, and includes by far the largest tonnage of all the conversion grades. The **basic** grade of iron is second in this country in quantity used and is rapidly overtaking Bessemer. It is designed especially for basic open hearth conversion. The **low-phosphorus** grade is naturally expensive, and therefore has not a very wide application. It is intended especially for conversion in acid open hearth furnaces, particularly for making highest grade steel castings and special steels. The **Thomas-Gilchrist** grade is intended only for the basic Bessemer process, which has now no application whatever in this country.

Classification of Steel Conversion Methods.—Of the four processes of conversion that have steel for their immediate object, viz.: Bessemer, Thomas-Gilchrist, acid open hearth and basic open hearth, the natural classification, based on method of operating, would be to distinguish the Bessemer and Thomas methods from the open hearth methods. The two former consist essentially of purifying pig iron in a nearly closed vessel by means of a blast of air. . For this reason, they are sometimes known as the **"pneumatic"** processes. The two latter consist essentially of melting the iron on a dish-shaped hearth and purifying it by means of iron ore thrown into the metallic bath.

But from the standpoint of the composition of the pig iron

to be used, it is more important to distinguish them according to what they can accomplish in the way of purification of the metal. For this reason, we shall consider them under the heads of **acid** and **basic** methods.

Acid Processes.—As we have seen already, the composition of the slag is a most potent factor in the purification of metals It is obvious, also, that it is impossible to control the composition of a slag if the furnace lining itself is practically a limitless reservoir of undesirable ingredients It is necessary, therefore, in order to produce certain results in purification, that the nature of the lining in the converting furnace should have careful consideration

We have seen repeatedly that silicon and manganese are natural slag-makers, and that when in the oxidized state they enter slags indiscriminately and regardless of slag composition. For the removal of such elements the composition of the slag need not be controlled within narrow limits and the least expensive type of furnace lining may be selected. Siliceous rocks or sands are the cheapest refractory materials obtainable and may be used as linings under the above conditions. Since silica is an acid radical, a siliceous lining will naturally permit a slag of acid predominance, and the process is therefore known as an "acid" process. The Bessemer and acid open hearth methods fall under this head.

Basic Processes.—When it is desirable or necessary to eliminate phosphorus and sulphur, as well as silicon and manganese, we know that a basic slag is necessary; and therefore in order to limit rigidly the supply of acid ingredients in the slag, the lining must be of basic material. The most successful basic refractory materials so far discovered are specially prepared magnesite and dolomite, which, on account of expense, are never used unless demanded by the composition of the iron to be converted. Under this head fall naturally the basic open hearth and the Thomas-Gilchrist, or, as it is generally called, the basic-Bessemer process

BESSEMER PROCESS.

The Bessemer process, which was invented by Bessemer in 1856, but not introduced here until 1867, consists essentially of pouring molten iron into a pear-shaped vessel, called a "converter," and blowing through it a blast of atmospheric air until it removes such impurities as the nature of the process permits, namely, silicon, manganese and carbon.

The converting vessel is a steel-riveted shell, some 20 feet high by 10 to 12 feet in diameter, reinforced and set on trunnions to facilitate rotation, having one end contracted to small diameter and the other closed by a detachable bottom. An air pipe leads from one trunnion which is hollow to an air box which is attached to the bottom, and which communicates with the tuyeres set in the bottom. The lining is at least a foot thick and may be made up of various materials, such as blocks of millstone grit or ground quartz mixed with fireclay, or silica bricks, etc., all of which are refractories of highly siliceous character. The lining is therefore denominated " acid."

Operation. —The method of operating a Bessemer converter is as follows: The vessel is turned down into the horizontal position and a charge of molten pig iron, usually 15 to 20 tons, from a blast furnace or cupola is poured in at the nose of the vessel. A blast of air, having a pressure of about 25 pounds per square inch, is turned into the vessel through the bottom tuyeres and the vessel is turned up into the vertical position. We then have the spectacle of some 20 tons of molten iron kept in violent ebullition by a constant current of air, whose force is sufficient to keep the fluid metal from running into the tuyere openings beneath it.

Elimination of Metalloids.—The immediate result of the action of the air upon the molten metal is the oxidation of the **silicon** and **manganese.** The resulting SiO_2 and MnO, together with some oxides of iron, immediately unite to form a slag. This period is characterized by a scanty flame coming from the nose of the vessel, since very little volatile gas is formed before the carbon is attacked The oxidation of silicon liberates a large quantity of heat and quickly brings the temperature up to the point of carbon ignition. If the temperature is allowed to rise too quickly, carbon will exert its preferential power over the oxygen, and the silicon and manganese will not be completely eliminated. The high temperature can be controlled by charging cold scrap into the converter during the blow, or by introducing steam with the blast.

The disadvantage of too much silicon and manganese is more

apparent here than in puddling, because the action is not so dependent upon the presence of slag or upon its composition. As before, they both represent so much waste to the purchaser, and do not compensate by reducing an equivalent of iron. The manganese, however, renders the slag more fluid, and at the same time displaces its equivalent of iron, thereby protecting the iron from excessive scorification. The silicon, on the other hand, does not, as in puddling, determine the quantity of slag absolutely, as the walls of the vessel naturally add to the quantity of silica, and consequently help in the carrying off of iron. The quantity of silicon affects, however, the length of blow, since the oxidation of carbon must not begin until the silicon is practically eliminated. It was formerly thought necessary to have 2 to 3 per cent. silicon in order to raise the temperature of the metal sufficiently, but with more rapid running it is found that 1 per cent. is ample.

Generally at the end of three or four minutes, the **carbon** begins to burn. The oxidation of carbon produces CO in the converter. As the CO reaches the outer air, it bursts into flame, forming CO_2. This period is marked by a long, brilliant flame at the nose of the vessel, and represents the maximum temperature attained. It has been calculated by J. W. Richards that the rise in temperature from that of the molten pig iron during a blow is nearly 600 degrees F., which brings the temperature of the product well above the melting point of wrought iron. If the flame shows a deficiency of heat at this period the temperature can be increased by tilting the vessel until some of the tuyeres are no longer covered by the metal. Unconsumed air, entering through these tuyeres, burns the CO within the converter and thereby raises the temperature of the vessel. Some iron is also oxidized, which increases the additional heat developed.

At the end of about ten minutes the carbon has been eliminated and the flame drops. The vessel is then turned down and a glowing mass of wrought iron, practically free from silicon and manganese, and having less than 0.1 per cent. carbon, is ready to pour into the ladle. A **recarburizer,** usually molten " spiegeleisen," is added to the metal in the converter before it is emptied. This recarburizer contains a definite percentage of carbon and manganese, which brings the composition of the resulting steel to the desired limits.

A modification of this process, carried on in a small vessel with side tuyeres and a low blast pressure is known as the **Tropenas process,** and is considerably used in making small steel castings.

Influence of Slag.—A study of the composition of Bessemer slags reveals the cause of the limitations of the process. Bessemer slags vary widely in composition, but the SiO_2 rarely falls much below 50 per cent., while it may rise to nearly 70 per cent. The bulk of the remainder is made up of the combined oxides of iron and manganese, the proportion of each being governed by the initial quantity of manganese. We have already seen that a blast furnace slag having as little as 30 per cent. SiO_2 is not sufficiently basic to carry off phosphorus, and even when the SiO_2 falls to 20 per cent , as in the puddling process, the extraction of phosphorus is far from complete. It is impossible, therefore, to expect any elimination of phosphorus in the Bessemer process. As a purifier from silicon, manganese and carbon it is unparalleled for speed and economy, but as an extractor of phosphorus or sulphur it has no claims whatever.

Composition of Bessemer Pig.—With regard to the limits of silicon, the Bessemer pig-iron maker is between two fires The silicon must not be so low that the heats will run cold, and thereby have no scrap-carrying power, or demand tilting, which wastefully oxidizes iron, nor should it be so high as to create an excessive quantity of slag or unduly prolong the time of blow. For these reasons, silicon should not fall much below 1 per cent. on the average or rise above 2 per cent., and manganese is only so much waste. As regards phosphorus and sulphur, since the Bessemer process is entirely unable to eliminate any portion of them, they will increase in percentage through concentration, because they remain unaffected while the quantity of metal is diminished by 8 to 10 per cent during conversion. They should therefore be carefully limited in the pig to about 90 per cent. of that allowable in the steel.

The final composition of Bessemer steel generally falls between these limits:

C Per cent.	Si. Per cent.	Mn. Per cent.	P. Per cent.	S. Per cent.	Fe. Per cent.
0 05-1.5	Tr -0.05	0 3-1.0	Below 0.1	Below 0.08	99 5-98 0

ACID OPEN HEARTH PROCESS.

The open hearth process, which was introduced in this country shortly after the Bessemer process, consists essentially of melting a mixture of pig iron and scrap on a dish-shaped hearth and oxidizing and removing certain impurities by means of iron ore and slags of suitable composition

The open hearth furnace consists of a large, horizontal rectangular chamber, built of very refractory bricks, properly bound by buckstays. The working chamber is usually about 30 x 15 feet, and holds 50 tons of molten metal At each end of the chamber are ports, which serve alternately as inlets for fuel gas and air which burn in the chamber, and as outlets for the products of combustion. Each gas and air port connects with an underground chamber filled with checker brickwork through which at one end the gas and air enter, and at the other the products of combustion escape. The heat extracted by the checkers from the waste gases serves, on reversing, to heat the entering gas and air. The furnace is known as the **Siemens regenerative furnace.** Its system of preheating the fuel and air easily permits the attainment of steel-melting temperature.

Character of Charge.—The bottom of the acid open hearth furnace is composed of very refractory silica sand which is fused in place. This fact determines the character of the slag and the limitations of the method. Upon this bottom of sand the metal is charged, melted and purified A striking difference between the pneumatic and regenerative systems of conversion lies in the character of the charge. In the Bessemer process we saw that it consisted entirely of molten pig iron In the open hearth process it may consist either of molten or of cold pig. Furthermore, the open hearth charge rarely consists wholly of pig iron, but a considerable quantity of steel and wrought iron scrap is charged also. The proportion of pig iron used in the charge of an acid furnace should be such that when the charge is melted, the silicon and manganese will have been eliminated and the bath will contain ½ to 1 per cent. of carbon Under ordinary conditions this requires 20 to 30 per cent. of pig and the remainder scrap.

Operation.—The method of operating an open hearth furnace has for its first step the charging of material into the fur-

nace. This was formerly done by hand by means of a long iron
" peel,' and this method is still used at some of the smaller
plants. It is slow, keeping the furnace open a long time and
increasing oxidation. The more usual method is to fill the furnace by means of specially designed boxes, handled by a **Wellman
charging machine.** A 50-ton heat may be charged in this way in
less than an hour. Rapid charging permits less loss of metalloids
and admits of using less pig iron in consequence. The rational
system of charging is to place the scrap on the hearth first and to
place the pig iron on top of the scrap. The covering of pig prevents the scrap from oxidation by the flame, and at the same time
it melts first and trickles down over the scrap, thereby carburizing
and dissolving it Hard firing should melt a heat completely in
three or four hours. When the metal is all under the slag and
shows by test to be hot and still high in carbon, lumps of ore are
thrown into the bath at intervals to oxidize the-carbon. Careful
watch of the carbon content tells when the heat has reached the
desired point, whereupon, if the metal is hot enough, it is tapped.
Ferromanganese and sometimes coke dust are thrown into the
ladle to ·recarburize and remanganize the metal to suitable composition.

Elimination of Metalloids.—As in the case of the acid Bessemer process, the **silicon** and **manganese** are eliminated first, their
elimination being generally accomplished by the time the fusion
is complete. As SiO_2 and MnO they unite with whatever oxides
of iron are present to form the slag. The slag floats on the top
of the metal and prevents, to a great extent, any further oxidation
by the flame. The elimination of the **carbon** is accomplished chiefly
by the ore which is thrown into the bath. The evolution of CO
agitates the bath, mixing it and facilitating the action. Here
again, carbon exercises its control over oxygen, and if the temperature is too high it will take it even from the slag. Both iron
and silicon may be reduced by it. The former action is a desirable
recovery of lost iron, but the latter introduces silicon and gives a
porous metal.

Composition of Pig.—The presence of **silicon** in pig for the
acid open hearth process has no advantages. Here, unlike the
Bessemer processes, there is no need of silicon to furnish heat,

because the bath is amply heated by fuel gas Silicon is so much
loss to the purchaser. Although a certain amount of slag is necessary as a covering to the bath, yet silicon is not absolutely necessary to its formation, as the sand of the hearth bottom could furnish enough for the purpose. Since all the silicon is usually
oxidized by the flame, it does not compensate in any way by reducing an equivalent of iron from the ore. In other words, it is
wholly undesirable, although in moderate quantities it does no
harm and tends to protect the hearth from excessive scorification
during the slag-forming period. A high percentage of silicon
would prolong the operation and cause large loss of iron by forming an excessive quantity of slag. It would compensate partially
for this loss, however, by reducing metal from the ore, since there
would be an unoxidized excess of silicon after melting, which
would be removed by the ore additions.

The presence of **manganese** also presents no advantages in this
process. It is oxidized and enters the slag, displacing an equivalent of iron, to be sure, but that does not affect the percentage
yield, since it was purchased as so much iron. It tends to make
the slag more fusible.

The oxidation of the residual **carbon** by means of the ore, results in the formation of CO and metallic iron, in accordance with
the reaction,

$$3Fe_x C + Fe_2O_3 = 3CO + Fe_x+_2,$$

thereby yielding an equivalent of iron to the bath of 3 1 pounds
Fe for each pound of C. The CO thus formed, bubbles up through
the slag and burns at the surface, yielding additional heat

The elimination of **phosphorus** and **sulphur** in the acid open
hearth process is subject to the same limitations as in the Bessemer
process. Acid open hearth slags usually have a silica content of
about 50 per cent., which is far too acid to permit the retention of
appreciable quantities of either element. Indeed, such a slag may
permit the transmission of noticeable quantities of sulphur from
the gases to the metal, and thereby give rise to an increase of
sulphur in the bath. All of the phosphorus and much of the
sulphur present in the ore incorporate themselves in the metal.
It is evident, therefore, that these elements should be strictly limited, not only in the pig iron, but also in the scrap, ore and fuel

Low Phosphorus Pig.—The limits which are imposed on the composition of pig which is to be converted by this process vary with the class of products desired. The limits of Si and Mn are naturally not so stringent as those of P and S, and as they both are completely eliminated, a given content of each is equally suitable for all purposes. With the P and S, however, the case is different. When it is desired to make metal of exceptional quality, such as government steel castings, ordnance, etc., which has specifications calling for P and S below 0.035 per cent., it is necessary to produce a pig having these elements not far above 0.030 per cent. Furthermore, it must be used only in connection with selected scrap and ore. This grade of pig has consequently received the name " low phosphorus " pig, although it must be equally low in sulphur. On the other hand, for structural steels, plate, etc., whose specifications are not too low, Bessemer pig, or a mixture of Bessemer pig with low phosphorus pig or with low phosphorus scrap, will bring the desired composition. It must be remembered that not only does all of the P and much of the S present in the bath congregate in the metal, but there is likely to be an apparent increase of them through concentration as the other metalloids are eliminated.

The final composition of acid open hearth steels generally falls between the following limits:

C.	Si	Mn.	P.	S.	Fe.
Per cent.	Per cent.	Per cent.	Per cent.	Per cent.	Per cent.
0.05-1.5	Tr.-0.3	0.3-0.7	0.03-0.10	0.03-0.08	99.5-98.0

BASIC OPEN HEARTH PROCESS.

This modification of the open hearth process was not introduced into this country until 1888. The furnace used in the process is identical with that used in its acid counterpart, except in regard to the lining below the slag line. As the name implies, the slag of this method is of basic predominance, and in order to maintain its basic character must be kept from contact with siliceous materials as much as possible. It is customary, therefore, to make the furnace bottom of basic material. The best construction consists of a thin lining of magnesite bricks, covered with a heavy layer of calcined magnesite, which is sintered with a small percentage of some fusible material such as a basic slag. Dolomite is sometimes used as a substitute for magnesite.

Character of Charge.—The charge of the basic furnace, like that of the acid, is made up of pig iron and scrap, except that the proportion is somewhat different. Owing to more oxidation during melting, a larger percentage of pig must be used. The pig usually ranges from 35 to 60 per cent. of the charge, and may be either molten or solid. Besides the metallic materials, a nonmetallic portion, known as the **"basic addition,"** is used. This consists usually of limestone or burnt lime, and iron ore. Its object is to neutralize the silica formed by the silicon oxidized during the melting period, and to prevent it from attacking the hearth. For this reason, the additions are charged first and spread over the furnace bottom, and upon them the metallic charge is placed. The quantity of basic material should be sufficient to neutralize completely the silica present and must therefore be varied according to the content of silicon in the pig used. The resulting slag should have about three times as much $CaO + MgO$ as SiO_2.

Elimination of Metalloids.—The oxidation in this process follows the usual order. During melting, **silicon, manganese** and **phosphorus** are oxidized in the order named. The resulting oxides trickle down with the molten iron and come in contact with the basic additions on the furnace bottom. The limestone has meanwhile been partially calcined and the lime unites to form the slag. The SiO_2 and the P_2O_5 are taken up readily by the oxides of calcium and iron, but, owing to the presence of such an excess of bases, the·MnO is not absorbed so readily, and therefore its elimination from the metal is not so rapid and complete as in the acid processes. The SiO_2, on the other hand, is snapped up eagerly and held so firmly by the very basic slag that it is not materially affected by the preference of oxygen for carbon, even at the highest temperatures. Approximately 60 per cent. of the **phosphorus** and 65 per cent. of the **carbon** on an average are oxidized during melting. The elimination of the phosphorus follows closely that of carbon, and with a properly basic slag the removal of phosphorus is practically complete by the time the heat is done. Under ordinary conditions a proper slag can usually be relied upon also to extract at least 50 per cent. of the **sulphur** originally present.

Influence of Slag.—The basic process was devised to enable the steel maker to cope with phosphorus and sulphur. As they both form acid radicals, it is only by means of extreme basicity that they can be securely held in the slag in the presence of the reducing influence of the carbon A slag cannot be an efficient carrier of phosphorus during the elimination of carbon at a high temperature if the acid content is much in excess of 25 per cent. of the slag. The percentage of P_2O_5 that can be carried, therefore, varies inversely as the percentage of SiO_2 present. The final composition of a good basic open hearth slag will be approximately as follows:

SiO_2	P_2O_5.	CaO + MgO	FeO	MnO
20 per cent.	5 per cent.	50 per cent.	15 per cent	10 per cent.

The necessity of limiting the SiO_2 in the slag so rigidly makes it imperative that silicon should be kept low in the pig iron. Since the percentage of SiO_2 must be so low in the slag, any increase of silicon in the pig means an increase of several times as much basic material, as well as the loss of an equivalent quantity of iron. The silicon in basic pig iron is, therefore, not only a dead loss to the purchaser, but entails the consumption of considerable other material which must also be purchased.

Composition of Pig.—The presence of **manganese** is of no especial advantage to basic pig iron. As a source of basicity its oxides are too unstable to be effective in eliminating phosphorus, although its affinity for sulphur may assist in the removal of that element Owing to the basicity of the slag, manganese is not so readily taken up by it in this process as in the acid processes, and its elimination may therefore be incomplete, which may result in too high manganese in the finished product after recarburization. It is best, therefore, to limit it also.

While the elimination of **phosphorus** in this process may be practically complete, yet it is accomplished only at considerable expenditure of basic reagents. We have seen that each pound of phosphorus needs several pounds of such reagents. It should therefore be kept low, as the cost of extraction multiplies rapidly. As a large portion of it is oxidized during melting, it does not yield its equivalent of iron to the bath. That portion, however, which is oxidized by the iron in accordance with the following reaction,

$$3P_2 + 5Fe_2O_3 = 5Fe_2 + 3P_2O_5,$$

yields 3 pounds of iron for each pound of P oxidized.

Owing to the fact that a considerable percentage of **sulphur** is taken up by a basic slag, it is not necessary that sulphur should be so strictly limited in basic pig as in acid irons. Usually 50 per cent. and often 75 per cent. of the sulphur in the pig can be extracted. A basic slag also prevents transmission of sulphur from the gases to the metal by assimilating it during its passage.

As in the other processes, **carbon** is the last element to be oxidized. The removal of the residual carbon is effected by the oxidizing effect of lumps of ore thrown into the bath. The reaction causes an evolution of CO, as in the following reaction,

$$3C + Fe_2O_3 = 3CO + 2Fe$$

which agitates the bath, exposing it to the action of the slag and flame. At the same time the carbon reduces an equivalent of iron equal to 3 pounds per pound of carbon oxidized.

The final composition of basic open hearth steel usually falls within these limits:

C.	Si	Mn	P.	S	Fe
Per cent	Per cent.	Per cent	Per cent	Per cent.	Per cent.
0.05-1 5	Tr.-0.2	0 3-0 7	0 005-0 05	0 02-0.06	99 5-98 0

THE BASIC BESSEMER PROCESS.

Operation.—The basic-Bessemer or Thomas-Gilchrist process, which was first applied in 1878, is a pneumatic process, carried on in a vessel differing from the Bessemer converter only in the character of its lining. The process was designed to supplement the power of the Bessemer process to eliminate rapidly and cheaply the silicon, manganese and carbon, by including the elimination of phosphorus and sulphur. As we have seen, this action demands a slag of pronounced basicity, which can be maintained only when the vessel lining is of basic character. The lining is usually made of dolomite which has been thoroughly calcined and mixed with a suitable binder, such as anhydrous tar. The vessel is charged with molten pig, turned up, and blown with an air blast similar to the Bessemer process.

Elimination of Metalloids.—The first elements to oxidize are, as usual, **silicon** and **manganese.** They unite to form a slag, which is of insufficient basicity to spare the dolomitic lining of the vessel. It is therefore customary to charge the converter before beginning the blow, with enough burnt lime to form a slag with all the other slag-making materials' present, which shall contain about 50 per cent. $CaO + MgO$. This serves to prevent excessive corrosion of the lining. Generally the removal of **silicon** and **manganese** is complete at the end of six to seven minutes, and the **carbon** is eliminated at the end of twelve to fifteen. While **phosphorus** is oxidized during these periods, yet the oxide is not stable in the presence of the active carbon and therefore practically none is eliminated until the carbon is gone. It usually takes three to five minutes additional blowing to remove the phosphorus. Incidentally, about half the **sulphur** is removed at the same time.

The final composition of steel made by this process is never as low in phosphorus as at the end of the blow, because the reactions accompanying recarburization generally reduce again a small portion of the oxidized phosphorus to the elemental condition.

Composition of Pig.—The presence of **silicon** in this process offers the same disadvantages that we have seen in the basic open hearth operation. The more silicon, the more basic additions are needed to neutralize it and give a slag sufficiently basic to hold phosphorus and sulphur. On the other hand a certain percentage of silicon is desired to raise the temperature of the heat to the ignition point of carbon.

Manganese, being of basic nature, serves to neutralize the silica and at the same time renders the slag more fusible.

The presence of **phosphorus** is the essential evil which makes the process necessary. The necessity for keeping silicon low in the pig in order that the slag may be sufficiently basic to take care of the phosphorus, throws upon phosphorus a duty which is performed by silicon in the acid converter. The phosphorus, since it is a heat producer second only to silicon, serves to supply the deficiency of heat which naturally results from the absence of considerable silicon.

As there is no compensatory reduction of iron, as in the case of purification by fixed oxygen, each element which is in excess of the requirements of the method is a total loss to the purchaser, even if it offers no especial disadvantages to the operation

The final composition of basic-Bessemer steel usually falls between these limits:

C. Per cent.	Si Per cent.	Mn. Per cent.	P. Per cent	S. Per cent	Fe. Per cent
0 05-1.5	Tr. 0.02	0.4-0.5	0.05-0.15	0 04-0.08	99 5-98 0

Chronology of Conversion Methods.—Although cast iron in the form of pig iron is now the starting point of the manufacture of all other ferrous products, its attainment to this distinction is of very recent date, and in metallurgical chronology, the blast furnace is a comparatively modern institution. The production of wrought iron direct from iron ores, and the manufacture of steel by causing bars of iron to absorb carbon through contact with charcoal at red heat, are industries whose origins are lost in antiquity. For many centuries wrought iron was produced by heating ore with charcoal and a small blast in low hearths, or hollows in the earth, known as the Catalan forge During the middle ages these hearths developed under ambitious managers into low masonry furnaces, with removable fronts for extracting the "loup" or lumps of iron. As furnaces became higher and reducing conditions stronger, the furnacemen were annoyed to find that a portion of their product was in a molten state, and that on solidifying it did not become tough and strong, but was comparatively brittle. This new material afterwards found a limited application in making castings, and the above method of making wrought iron continued for some time longer. Subsequently methods were devised for refining the cast iron, which eventually displaced the production of wrought iron direct from the ores. The earlier methods of refining consisted in melting the cast iron in small quantities in small rectangular hearths by means of charcoal and a blast which oxidized the metalloids. The method was wasteful and the fuel expensive. The puddling process, which was introduced by Henry Cort in 1784, was considered a great improvement, because the metal did not come in contact with the fuel and an inferior grade could be used in consequence. But the hearth was made of sand, and as there was no compensatory reduction, the losses were large. The present modification was introduced by Hall in the thirties.

It was not until 1856 that a method was developed for using pig iron in producing steel direct. In that year Bessemer discovered, in England, that the metalloids can be eliminated from molten pig iron simply by blowing through it a current of atmospheric air. The discovery was the result of an endeavor to find a shorter method of making wrought iron. The product, while ostensibly wrought iron, would not roll without crumbling, owing to the admixture of oxidized particles of iron. It was observed by Mushet that the addition of a small percentage of metallic manganese removed the oxygen and gave the metal the malleable quality which is so familiar in mild steel. The basic modification of this process was also brought out in England during the seventies.

Meanwhile, during the sixties, the acid open hearth method was developed by Martin of France, who adapted the regenerative furnace of Siemens to melting pig iron and scrap iron, thereby introducing a new method of steel making, still sometimes known as the Siemens-Martin process, although the process has since been somewhat modified by the use of iron ore. The use of a basic hearth was adopted on the Continent some years later.

NON-CONVERSION IRONS.

Of the irons which do not undergo complete conversion during preparation for further use we may distinguish two classes, those which undergo a partial change in composition in making malleable or toughened castings, and those which are used practically unchanged in making gray iron castings. The latter class includes all the so-called foundry grades of iron.

GRAY IRON CASTINGS.

The first use of gray pig iron to make castings is of uncertain date, but probably was previous to 1500 A. D., since cast iron cannon are known to have been in use at about that time. Before its introduction, all articles of iron were made of wrought iron by the laborious process of forging and welding. The building up of complicated shapes in that way is slow and expensive, and therefore the invention of a modification of iron which was suitable for many purposes and yet could be melted and poured directly into complicated shapes was very welcome. In spite of the cheapened methods of producing wrought iron and steel, and the recent introduction of steel castings, iron castings still fill a very wide field of usefulness, and must continue to do so, not only on account of their low cost of production, but also because the metal is better suited to some uses than either wrought iron or steel.

Grading Foundry Iron.—It was long the custom to grade pig iron for its various uses according to the character of its fractured surface. The different grades were usually designated by numerals, as follows:

1 X.—Open grain, having large crystals of graphite to the very edge.

2 X.—Slightly closer grain with a markedly closer border.

2 Plain.—Closer than 2 X, especially toward the bottom.

3.—Uniformly closer than 2 Plain.

It is now fast becoming customary to buy iron on its analysis instead of its fracture, and the interpretation of these grade numbers in terms of composition is, according to the Warwick Iron Co., as follows:

"Analyses of Pig Iron," S. R. Church, p. 89.

Grade	Si.	T. C	C. C.	Mn.	P.	S
1 X. ...	2 0-3 0	3 5-4 0	0.1-0.3	0.4-0.6	0 4-0 5	0.01-0 03
2 X.	2 0-3.0	3.5-4.0	0 2-0 4	0.4-0.6	0.4-0.5	0.02-0.04
2 plain	2.0-3 0	3.5-4.0	0.2-0.5	0.4-0 6	0.4-0.5	0 02-0 06
3	1.0-1.75	....	....	0.4-0.6	0.4-0.5	0 04-0.08

A more elaborate classification which has grown up in the South to meet local conditions is illustrated by the analytical limits of the Alabama Consolidated Coal and Iron Co.'s products.

Ibid, p. 17.

Grade.	Si.	G C.	C. C.	Mn.	P	S.
Silvery	4.0 -6.0	2.25-3.5	0.35-0.50	0 75-1 0	0.17-0.30	0.015-0 030
No. 2 soft	3.0 -4.5	2.25-3 5	0.35-0.50	0 75-1.25	0.20-0.30	0.020-0 030
No. 1 soft	3 0 -3.8	2.25-3.5	0 35-0.50	0 75-1.25	0.20-0.30	0.020-0.025
No. 1 foundry	2.5 -2.75	2.25-3.5	0.35-0 50	1.10	0.30	0.030
No. 2 foundry	2 25-2.5	2.25-3 5	0.35-0.50	1.25	0.35	0.035
No. 3 foundry	2.0 -2.25	2.25-3.5	0.35-0.50	1.30	0.40	0 040
No. 4 foundry	1.75-2.0	2.25-3.5	0 35-0 50	1.35	0.45	0.040

The physical character of pig iron is dependent mainly upon the condition of the carbon, and the most potent factor in determining the condition of the carbon is the quantity of silicon present. Since silicon is not the only controlling influence, it frequently happens that it does not exert its normal influence on the carbon, whereupon the fracture ceases to be a reliable guide to the quantity of silicon and hence to the quality of casting that will result It is now more usual, therefore, to grade the iron according to its silicon content rather than according to the appearance of the fracture.

A given make of iron frequently fails to run uniformly in quality and the consequent variation might cause trouble for a given class of work To get best results, therefore, it is customary to mix several makes or " brands " of a given grade, rather than to risk all results on a single brand with its liability to variation. Accidents are less likely, however, when iron is bought strictly on analysis.

Properties of Foundry Irons.—The wide variation in the properties of cast iron permits its application to a great variety of uses. The particular qualities which adapt it to making castings are its ready fusibility and low shrinkage These qualities are directly attributable to the quantity and condition of the carbon present. As previously stated, the quantity of carbon in pig iron is fairly constant, but its condition in castings is subject to wide variations. There are four chief factors which affect the condition of the carbon, viz.: The quantity of carbon present; the initial temperature of the metal, the rate of cooling; the presence of other elements. The effects of these various factors upon the properties of cast iron may be conveniently summarized in the following table -

	Softness.	Hardness.	Strength	Weakness.	Brittleness	Shrinkage.	Chill.	Fluidity.	Sluggishness.	Blowholes	G. C.	C. C.	Lowers M. pt.	Raises M. pt.
Total C.	*			*				*			*		*	
Comb C.			O	[]					*				*	
Graph C	*		O	[]			*							*
Si	O	[]	O	[]			*							*
Mn	O	[]	O	[]	[]	*		*				*		
P						*		*			*			
S			*	*	*					*				*
Slow cooling.	*			*						*				*
Rapid cooling	*	*		*	*	*							*	

O = small percentages.
[ ] = large percentages.
* = all percentages.

PURCHASE SPECIFICATIONS FOR FOUNDRY IRONS.

The standard purchase specifications for pig iron adopted by the American Society for Testing Materials is as follows:

1. All purchases to be made on analysis.

2. Each carload to be considered a unit, one pig from each four tons to constitute a sample. Drillings taken from fractures of pigs to fairly represent the pig and an equal quantity from each pig mixed thoroughly and ground before analysing. In case of dispute an independent analysis to be made on one pig for each two tons, the cost to be borne by the party in error.

Soc.
Ist.,
103

3. All contracts, unless otherwise agreed, to allow a variation of 10 per cent Si either way and 0.02 S above the standard for the given grade A deficiency of silicon between 10 per cent. and 20 per cent., subject to 4 per cent. deduction in price.

4. In absence of other agreements the following analyses represent standard grades of foundry irons:

	Per cent. S_i.	Volumetric. Per cent S.	Gravimetric. Per cent. S
No. 1	2.75	0.035	0.045
No. 2	2.25	0.045	0.055
No. 3	1.75	0.055	0.065
No. 4	1.25	0.065	0.075

In general it may be stated that for average foundry irons the following rules hold good:

To increase strength of castings: Decrease phosphorus and lessen graphite by decreasing silicon, thereby allowing more combined carbon. The manganese may also be increased and the castings cooled more rapidly.

To decrease hardness and shrinkage of castings: Decrease sulphur and combined carbon and increase the quantity of graphite [phur and combined carbon and increase the quantity of graphite] through the addition of silicon or by cooling more slowly.

To prevent chilling of castings: Decrease sulphur and manganese and increase silicon and slow cooling.

To prevent blowholes in castings: Decrease sulphur and increase manganese and silicon.

To prevent kish: Decrease the percentage of carbon by adding scrap to the cupola.

The addition of small quantities of ground ferromanganese

or 50 per cent. ferrosilicon in the ladle gives generally beneficial results. It deoxidizes the metal, thereby softening and strengthening it, decreasing shrinkage and making clean castings without materially altering the composition of the metal. Iron Age, Nov. 1, 1906, p. 1140

Pig iron which shows no tendency to vicious properties such as hardness, weakness, shrinkage, etc., is sometimes termed "neutral" pig.

In order to illustrate the type of metal that is suited to castings of different degrees of hardness, the specifications of the Case Threshing Machine Company will serve as an example:

	Soft castings, pulleys or small castings	Medium castings, cylinders, gears, pinions, etc.	Hard castings, valves and H P. cylinders.	
Si	2.20-2.80	1.40-2.00	1.20-1 60	
S	below 0 085	below 0 085	below 0 095	Iron Age, Sept. 29, 1898, pp. 4, 5.
P	below 0 70	below 0.70	below 0 70	
Mn	below 0 70	below 0.70	below 0 70	
Tensile strength per square inch	18,000 lb.	20,000 lb	22,000 lb	
Transverse strength per square inch	2,000 lb.	2,200 lb.	2,400 lb	
Deflection, not less than	0 10 in.	0 09 in	0.08 in.	
Shrinkage, per foot, not over	0 127 in	0 136 in	0 148 in.	
Chill, not over	0.05 in.	0.15 in	0 25 in.	

The decrease of silicon is accompanied by an increase of combined carbon, which effects the increase of strength and stiffness of the metal. As the shrinkage generally varies with the hardness of cast iron, it becomes a valuable indication of the quality of iron, since it is measured with comparatively little trouble. Shrinkage is usually reckoned to be about ⅛ inch to a foot, but it is variable and depends upon the silicon content and area of section, as indicated by the following table:

Si.	½ x ½ in.	1 x 1 in.	1 x 2 in	2 x 2 in	3 x 3 in.	4 x 4 in.	
1 per cent	0.183	0.158	0.146	0.130	0.113	0.102	
2 per cent	0.159	0.133	0.121	0.104	0.083	0.074	Keep's "Cast Iron," p. 155.
3 per cent	0.135	0.108	0.095	0.077	0.059	0.045	
Ratio Surface/Volume	0.125	0.250	0.333	0.500	0.750	1.000	

The class of iron that is suited to certain uses may be illustrated by the composition of the samples used by the Committee on Standardizing Testing of Cast Iron.

Journal
Am. Fd'ymen
Assoc.,
X, part II,
p. 25.

Class of work	Si.	P.	S.
Novelties	4.19	1.236	0.080
Stove plate	3.19	1.160	0.084
Cylinders	2.40	0.839	0.084
Light machinery	2.04	0.578	0.044
Heavy machinery	1.96	0.522	0.081
Dynamo frames	1.93	0.405	0.042
Ingot molds	1.67	0.095	0.032
Car wheels	0.97	0.301	0.060
Chilled roll	0.85	0.482	0.070
Sand roll	0.72	0.454	0.070

Effects of Size of Castings.—As the size of a casting exerts a marked influence upon its rate of cooling, it follows that composition should vary with the size of casting. Since a thin section cools rapidly, it should be made of iron which is very fluid and has no tendency to chill. On the other hand, a large mass of metal which cools slowly should not be allowed to form large crystals of graphite in its interior. In general it may be said that small castings should be high in Si and P, while large ones should be high in Mn and S. The following limits of composition, based on size of castings, have been suggested:

Iron Age,
Feb. 15, 1906,
p. 589

Thickness of section.	Si.	P.	Mn.	S
Under ¼ inch thick	3.25	1.00	0.40	0.025
¼ to ½ inch thick	2.75	0.80	0.40	0.040
½ to ¾ inch thick	2.50	0.75	0.50	0.050
¾ to 1 inch thick	2.00	0.70	0.60	0.060
1 to 1½ inches thick	1.75	0.65	0.70	0.070
1½ to 2 inches thick	1.50	0.60	0.80	0.080
2 to 2½ inches thick	1.25	0.55	0.90	0.090
2½ to 3 inches thick	1.00	0.50	1.00	0.100

As a rule, cast iron is not required to stand a transverse strain of over 2500 pounds per square inch. Generally small sections are not made for strength, and are therefore ample for the demands. Unsymmetrical castings may be benefited by exposing the larger sections to cool first.

Effect of Shape of Castings.—Since cast iron during solidification tends to build up a crystalline structure which grows in directions perpendicular to the cooling surfaces it follows that a plane of weakness will develop at every sharp angle, thus:

For this reason care should be exercised to have no abrupt changes of direction in patterns, but well rounded corners, thus:

in order that no distinct plane of cleavage may develop during solidification.

Remelting.—In making castings from any metal it is necessary that the metal should be in a thoroughly liquified condition in order that it may reach all parts of the mould uniformly. If a blast furnace could be run with such regularity as to turn out iron of a tolerably constant composition, it would be economy to run foundries in connection with blast furnace plants. Certain plain, heavy castings are made at every furnace plant at times, but no foundries in this country are run on direct blast furnace metal. In England the attempt has been made to operate in this manner to produce heavy work, by using carefully selected materials and carefully regulated temperatures in the furnace. In this country, however, it is the universal custom to remelt the pig iron in cupola furnaces and pour the remelt into the moulds.

Operation of Cupola.—The cupola furnace is a cylindrical, riveted, plate affair, lined with firebricks and pierced near the bottom for tuyeres. It is usually about 10 feet high and varies in diameter up to 8 or 10 feet, according to capacity required. The cupola charge usually consists of a bed of coke of 1000 to 2000 pounds, according to the size of the cupola. Upon this 3000 to 6000 pounds of iron are charged. Subsequent charges are usually smaller, ranging from 400 to 4000 pounds of iron and a gradually decreasing proportion of fuel. The usual fuel ratios are 1 pound of coke to 7 to 10 pounds of iron. A small percentage of limestone is generally charged to flux the fuel ash, and any sand which may adhere to the pigs. A vigorous combustion is maintained by a blast of air at atmospheric temperature, forced through the tuyeres at a low pressure, usually 10 to 20 ounces per square inch. The blast may be propelled either by a fan blower, such as the Sturtevant, or a positive blower like the Connorsville. Ordinarily the positive blower will furnish blast at a higher pressure with less expenditure of power per ton than the fan blower, and the rate of melting is proportionately higher. At low pressures, however, the fan blower is more flexible and furnishes a more uniform blast.

Rate of Melting.—The melting zone of the cupola depends for its position upon the force of the blast. A light blast will burn the coke near the tuyeres. A high pressure blast drives the zone of combustion higher. Usually a pressure which melts at about 18 inches above the tuyeres is considered best.

The rate of melting will depend upon the fuel ratio and the rate of combustion. The rate of combustion is accelerated by pressure of blast but is always in proportion to the quantity of air blown. The quantity of air varies as the square root of the pressure, and bears a tolerably definite ratio to the quantity of fuel. Generally the ratio $\frac{CO}{CO_2}$ in the escaping gases is about 1, from which it is easy to see that each pound of coke needs 7.3 pounds, or 105 cubic feet of air at 60 degrees F. From this it is evident that

A melting ratio of 6 requires 39,200 cubic feet air per ton pig.
A melting ratio of 7 requires 33,600 cubic feet air per ton pig.
A melting ratio of 8 requires 29,400 cubic feet air per ton pig.
A melting ratio of 9 requires 26 100 cubic feet air per ton pig.
A melting ratio of 10 requires 23,500 cubic feet air per ton pig.

The rate of melting also varies with the size of the cupola. The pressure required, and, consequently, the rate of melting, will rise as the diameter of the cupola increases. Generally it may be said that a

Age,
1905,
451
30-inch cupola requires 8 ounces pressure and will melt 2- 3 tons per hour.
45-inch cupola requires 10 ounces pressure and will melt 6- 7 tons per hour.
60-inch cupola requires 12 ounces pressure and will melt 10-12 tons per hour
72-inch cupola requires 14 ounces pressure and will melt 16-18 tons per hour.
84-inch cupola requires 16 ounces pressure and will melt 21-24 tons per hour.

It appears from this, that each square foot of cupola area melts 0.5 to 0.6 tons pig per hour, and the pressure in ounces should approximate the square root of three times the cupola diameter in inches. Usually 0.00045 H. P. will deliver 1 cubic foot of air per minute at 1 ounce pressure. With centrifugal fans, the melting capacity of the cupola is nearly proportional to the speed of the fan, the pressure is nearly proportional to the square of the speed, and the H. P. used is approximately proportional to the cube of the speed. Iron Age, June 23, 1904, p. 12

The use of a central tuyere in the bottom of a cupola in conjunction with the usual tuyeres tends to make combustion more complete, thereby saving fuel and increasing the melting rate, but it is difficult to maintain, owing to its exposed position. Inst Jour, 1897, I, p. 272.

Changes in Composition.—While the remelting of iron in the cupola is not intended to affect its composition and properties, as a matter of fact it rarely yields an unchanged metal and sometimes the alteration may be considerable.

The factors which conspire to alter the composition of the iron during melting are two: the blast and the fuel. The former tends to remove some elements through oxidation, while the latter tends to add some through absorption. Silicon and manganese, being readily oxidized, are invariably lessened during remelting, the degree of change being governed by the time that the metal is exposed to the action of the blast Sulphur and phosphorus are usually absorbed from the fuel, the amount depending upon the quantity of fuel and its purity Carbon in the metal may increase or decrease according to the influence to which the metal is most exposed. Eng'r News, XLVIII, p 46.

The actions of **silicon** and **manganese** during remelting follow each other closely and vary according to two conditions, viz., the quantity of the elements present and the quantity of oxygen to which they are exposed If the percentage of these elements is high, not only is the quantity lost larger but the proportion is higher. For example, when silicon is as high as 3 per cent the loss is generally more than 10 per cent. of that present, while when only $\frac{1}{2}$ per cent. is present, the loss may be scarcely noticeable. Both iron and carbon have a protecting influence upon silicon when it is in small quantities.

The greater the quantity of blast and the longer the metal is exposed to it the more oxidation will take place, and the greater will be the loss of silicon and manganese. For this reason, the loss by oxidation is generally greater in larger cupolas. The following statement represents average results:

Cupola diameter.		Silicon lost	Manganese lost
Under 40 feet. ..	.. ..	10 per cent	15 per cent.
40 to 60 feet.	....	15 per cent.	20 per cent
Over 60 feet .	....	20 per cent	25 per cent

In addition to loss by oxidation manganese may be carried off by combining with sulphur to form MnS, which enters the slag. The loss of Mn is generally greater in high sulphur irons. When the presence of manganese is desirable, and excessive sulphur is present the addition of ½ per cent. of ferromanganese will counteract this tendency.

The absorption of **phosphorus** by pig iron during remelting in cupolas is never very great. The tendency is present, but the action is incomplete, and, as a rule, the percentage of phosphorus is fairly constant during remelting.

With **sulphur,** however, the action is more positive. The source of sulphur is the fuel, and the tendency to absorption is proportional to the quantity of sulphur present. This tendency is counteracted by three influences, namely, the quantity of limestone used as flux, the temperature of the cupola, and the quantity of manganese present. A highly sulphurous iron melted in the presence of considerable metallic manganese may even lose sulphur. The use of manganese ore for this purpose is not efficacious, because sulphur does not unite readily with oxides of manganese, and the cupola is unable to reduce the metal. The use of ferromanganese in the ladle is said to remove 50 per cent. of the sulphur present.

The action of **carbon** during remelting also depend upon conditions, the oxygen of the blast tending to eliminate it and the coke tending to add it. As the iron passes the tuyeres, the tendency of the blast is to oxidize the carbon. This tendency is more marked if the carbon is plentiful, the exposure severe and the protection of silicon and carbon insufficient. On the other hand, long contact of molten iron with coke at high temperatures facilitates the absorption of carbon and the quantity may increase. In general, it may be stated that,

Much carbon, little Si and Mn, much blast and low fuel quantity tend toward decrease of carbon content, while

Low carbon, much Si and Mn, light blast and much fuel tend to increase carbon content.

Effect of Fuel.—Aside from its impurities, the quantity of fuel used in remelting pig iron affects the composition of castings through its calorific properties. Rapidity of melting tends to lessen the period of oxidation, and gives soft castings. Good and sufficient fuel brings down the iron hot and fluid, so that impurities separate well and the moulds are well filled. For these reasons, selected fuel is generally used in the form of 72-hour beehive coke. The use of by-product coke is not unusual, comparison shows that it is capable of melting iron satisfactorily. Moreover, as its method of manufacture involves the presence of a chemist, it is possible to get better information concerning it than in the case of beehive coke. Too little fuel causes slow melting and cold iron, permitting much oxidation and absorption of sulphur, which leads to hard castings. Good fluxes are necessary to free the iron of its impurities and make clean and smooth castings.

Iron Age,
Jan 11, 190
p. 16

Effect of Tuyeres.—The position of the tuyeres of the cupola is not without its effect upon remelting. For heavy work it is necessary to have them high, in order that the space below may be a sufficient reservoir for the quantity of iron needed for a large casting. This permits long contact with the fuel, and hence absorption of both carbon and sulphur, which is not detrimental to heavy work.

A cupola with low tuyeres must be drawn frequently and hence is suitable only for light work, and when low carbon and sulphur are required.

Action of White Iron.—Gray iron melts more slowly than white iron, and needs a higher temperature, owing to the necessity of the recombining of the graphite, but when it is melted it is usually more fluid than white iron. White iron becomes somewhat pasty on cooling, and entangles gases, causing blowholes. White iron melts as low as 2000 degrees F., while gray iron may require more than 2200 degrees F. In general, it may be said that the higher the total carbon, the lower will be the melting point, and the more fluid the iron. Gray iron throws off a layer of graphite on solidification, which covers the surface of castings, and prevents them sticking to the sand, thereby giving them a smooth appearance. Owing to their different rates of melting, gray and white iron do not give uniform results when melted together in the cupola Usually the white iron comes down first and is followed by the gray iron.

CHILLED CASTINGS.

For certain purposes, however, such as the formation of hard surfaces, white iron is very desirable. Castings with a surface layer of white iron backed by gray iron are known as **chilled castings.** They are made from irons low in silicon, in moulds having a metallic surface which corresponds to the surface to be chilled. The metallic surface conducts the heat away from the casting so much more rapidly than sand that there is not sufficient time for the separation of graphite It is always necessary to use an iron low in silicon in order that there may not be too great an initial tendency to form graphite. It is difficult to chill irons with over 2 per cent. of silicon The presence of sulphur and oxygen increases the chilling tendency. It is said that 0.01 per cent. sulphur will neutralize the softening effect of 0.1 per cent. silicon The addition in the cupola of steel scrap, chilled or white iron or manganese has the effect of deepening the chill A manganese chill, however, is undesirable on account of a tendency to "spall." With average sulphur and manganese low, silicon should give the following results:

> 1.00 per cent. Si gives ⅛ inch depth of chill.
> 0.75 per cent. Si gives ¼ inch depth of chill.
> 0.50 per cent. Si gives ½ inch depth of chill.
> 0.40 per cent. Si gives 1 inch depth of chill.
> 0.30 per cent. Si gives 1½ inch depth of chill.

If sulphur and manganese are high, the chill for a given silicon will be correspondingly deeper. The addition of small quantities of ferromanganese to the ladle will lessen the depth of chill, through the removal of sulphur and oxygen.

Non-chilling irons may be used in making chilled castings, either by mixture with chilling irons, in the cupola, or by melting direct in the air furnace, by which the composition may be modified through oxidation until it acquires the chilling tendency. For such castings the air furnace has some advantage over the cupola; the content of silicon, manganese and carbon may be regulated more exactly; the tendency to absorb sulphur as silicon decreases is less, since the metal is not in contact with the fuel;

and larger quantities of metal may be prepared simultaneously. This method is usually employed in producing chilled rolls. They are usually cast on end with the body in chills and the necks in sand. A heavy sinkhead serves to supply additional metal during contraction. Iron Age, Apr. 23, 1903, p. 2

Charcoal pig usually shows great strength and a tendency to chill, therefore it is in demand for making carwheels, which require great strength combined with a very hard, durable tread. The composition recommended by the American Society for Testing Materials is as follows: Tr. Am. Soc., M. E., XX, p. 615.

Total carbon....................... ....3.50 per cent.
Si .   ...0.70 per cent.
Mn ....0.40 per cent
P ....,.............. ....0.50 per cent
S ....0.08 per cent.

Iron Age, July 20, 1905, p. 162.

TOUGHENED CASTINGS.

Gray iron castings rarely give over 2000 pounds per square inch transverse strength, or 20,000 pounds tensile strength. Owing to demands for stronger castings for use in high pressure cylinders, pumps, etc., the practice of melting gray iron with varying percentages of mild steel or wrought scrap has been adopted. The resulting metal goes under the name of **"semi-steel,"** although from its former use in making gun carriages it is sometimes known as **"gun-iron"** The use of 20 to 30 per cent. scrap will generally increase the strength by at least 30 per cent. through decreasing graphitic carbon, and strengths of over 3000 transverse and 30,000 tensile are the rule. Such metal can successfully replace many steel castings. Care must be exercised in the use of scrap, however. Steel increases shrinkage; hard steel makes hard spots, wrought iron increases porosity; thick castings will stand more than thin ones. The effect of various additions is illustrated in the following table: Eng'r News XLIX, p. 309

Per cent steel.	Si	Mn.	P.	S.	C C.	T. C.	Trans. str.	Per ct incr.	Ten str.	Per ct. incr.
0........	1.96	0.44	0.446	0.104	0.63	3.18	2,230		21,950	
25........	1.50	0.33	0.532	0.065	0.64	3.44	2,840	27	30,500	40
0.	1.76	0.53	0.488	0.002	0.51	3.12	2,440		22,180	
25........	1.83	0.55	0.610	0.100	0.51	2.44	3,280	35	36,860	65
12½.. ...	2.16	0.20	0.315	0.060	1.06	2.30	2,670		26,310	
25........	2.36	0.24	0.327	0.064	1.08	2.15	3,200	20	31,560	20
0........	2.35	0.56	0.515	0.061	0.54	3.40	2,200		21,990	
37½.......	1.97	0.48	0.470	0.093	0.57	2.83	3,050	35	32,530	50

Iron Age, June 19, 1902. p. 27.

EFFECTS OF MOULDING

The production of good castings is very dependent, also, upon the character of the moulding sand and the arrangement of the moulds, particularly with respect to the proper gateing and distribution of the metal

Iron Age,
Feb. 13, 1896,
p. 415

The essential qualities of a good moulding sand are toughness and porosity. The sand when dampened should cohere on pressure and retain imprints distinctly, yet should be porous enough to allow escape of hot air, steam and gases.

The essential constituents of moulding sand are: Free silica or sand, and silicate of alumina or clay. The former gives porosity; the latter strength. Porosity is diminished by the presence of clay, by the fineness of the sand and by the regularity of the shape of the grains. The finer the sand the better the surface of the castings. The strength of the sand should vary with the kind of work. The sand depends for its strength upon the properties of the clay or " bond," the shape of the sand grains and the thoroughness of mixing. Strength may be supplemented by using binding compounds, or by the use of nails or " jaggers."

Iron Age,
Mar. 15, 1906,
p. 951.

Moulding Sand.—Good moulding sand will usually be found to have a composition between the following limits:

SiO_2 ...... 75-85 per cent ⎱ of which 60-70 per cent. = free SiO_2, and 20 30
Al_2O_3 .... 7-10 per cent. ⎰   per cent. = clay bond.
Fe_2O_3 below . 6 per cent.
CaO below ..... 2 per cent
Alkalis below ...¼ per cent.

The sum of the last three items should never exceed 7 per cent., as they would seriously interfere with the refractoriness of the sand. The higher the percentages of pure quartz and pure clay, and the coarser the sand, the greater will be the refractoriness Large castings require sand of greater refractoriness than small ones

Moulds.—Sand moulds may be roughly classified into green sand, dry sand and loam moulds.

Green sand moulds consist of moist sand, rammed around a pattern in wood or iron " flasks," which receive the metal direct.

Jour. Ass'n
Eng'r Soc.,
XXII, p. 209

Dry sand moulds differ from green sand only in the fact that the mould is dried at a moderate temperature for 24 hours before use. They are much firmer than green sand, and hence are used for larger castings. They also permit the use of higher temperatures, which is favorable to fluidity and hence cleaner metal.

Loam moulds are used only for very heavy work. Usually no pattern is used, but the loam is shaped up roughly over a backing of brick on the foundry floor, then finished and faced with finer material. They are generally reinforced by binders and jaggers and are dried thoroughly before use.

Cores are used when holes are desired in a casting, and are usually made of sand with some form of extra " binder," such as molasses, flour, resin, beer, gluten or some of the many " core compounds" on the market. They are shaped and baked before putting in place.

Iron Age, Feb 9, 1905, p. 483.

Gateing.—The proper gateing of a mould is quite as important to good castings as the character of the sand. The inlets should be ample to admit the metal rapidly and should be symmetrically placed. If more than one is necessary, they should be so distributed that the metal will fill evenly and freely to prevent cold-shuts. For large castings, the gate should be provided with a "riser" which will hold reserve of molten metal to feed the contraction space of the casting.

Cooling.—As soon as the metal has solidified, the castings should be removed from the moulds in order that they may cool naturally. The slow cooling in sand is equivalent to annealing. It will give a large, open grain, accompanied by weakness.

Cleaning.—Small castings may be cleaned from adhering sand by rotating in a tumble barrel. Large castings may be cleaned by means of a sand blast. Pickling in acids is also practised, although it has a tendency to weaken the metal.

Iron Age, March 12, 1896 p. 640

MALLEABLE CASTINGS.

The production of malleableized castings consists essentially of first producing a hard, brittle casting of white iron and subsequently rendering it tough and malleable by further treatment. This result may be attained in two ways, either by heating in the presence of oxidizing agents, such as iron ore, thereby removing the carbon wholly or in part, or by heating without oxidizing, thereby converting the combined carbon largely into uncombined, when it exerts but little influence upon the metal. This process was first described by Reaumur in 1722, but was not introduced into this country until a century later. The first works in America was established at Newark, N. J., in 1826. The use of malleableized castings has found wide application and there are many plants now devoted to the industry.

Jour Franklin
Inst.,
1899, II,
p. 134

Melting.—The metal for malleableized castings may be made either in the cupola, the open hearth or the air furnace, so long as the molten metal is of the right composition. The most. generally used form of furnace, however, is the air furnace. It consists of a rectangular chamber with a fireplace at one end, and the chimney flue at the other. In general, it differs but little from a puddling furnace, except in the character of its bottom, which is made of bricks with a sand covering. A charge of pig is melted down and rabbled, and the slag skimmed off. The oxidizing action during melting and rabbling permits the oxidation of the silicon and some of the carbon. The rabbling and skimming is repeated at intervals until a test shows that the metal is completely white. A test is taken for temperature and fluidity, then the metal is tapped into ladles and poured into green sand moulds, just as in the case of gray iron castings. Care should be taken not to chill the surfaces, as a chill remains white during annealing, and affects the strength of the casting.

Annealing.—The castings are freed from sand by tumbling, or by pickling in dilute sulphuric acid, and are then ready for annealing. They are packed in fine ore, or roll scale in cast iron boxes whose joints are luted with clay to exclude air. The annealing takes place in ovens which must be heated above redness. The time necessary decreases with increased temperature, the minimum being 60 hours for light castings and 72 for heavy. The average temperature is from 1600-1700 degrees F

Am. Soc.
Test Mat.,
III, p. 204

Rationale.—During the period of annealing CO is given off at the joints of the boxes. It burns with a blue flame and the iron content of the ore rises. This shows that a reaction takes place between the oxygen of the ore or scale and the carbon of the metal, whereby CO is given off. The reason for this action becomes evident when we recall the constitution of the metal. Carbon in iron may assume four different forms. When the metal is high in silicon a small portion of it unites with the iron, but the greater part exists as **graphite.** When silicon is wholly or nearly absent the carbon exists in the " combined " condition, of which part exists as a definite carbide or **cementite,** having the formula, Fe_3C, and the remainder unites with the excess of iron in what is known as the hardening condition or **martensite.** Under the influence of heat, this combined carbon separates and forms black spots, scattered throughout the metal. These were named **temper carbon** by Ledebur. Temper carbon is black, solid carbon, similar to graphite in nature, but differing from it in the fact that it is amorphous and will recombine with iron at temperatures as low as 1700 degrees F. It does not separate at any temperature if the total carbon is less than 0.9 per cent. This separated carbon, which exists near the surface of castings is free to unite with the oxygen of the ore or scales which surround it, and CO is the result. Carbon from the interior of the casting diffuses toward the surface, where it is oxidized in turn. In this way the quantity of carbon may be gradually decreased, as shown by the following table.

Time.	Per cent. total carbon.	Per cent hardening C.	Per cent cementite.	Per cent. temper C.	
Before heating	3.338	0.741	2.597	....	
Fourth day	3.061	0.815	2.246	....	Inst. Jour,
Fifth day	2.932	0.859	2.073	....	II, 1897,
Sixth day	2.888	0.835	1.874	0.179	p. 460.
Seventh day	2.098	0.631	0.430	1.037	
Eighth day	1.570	0.245	0.492	0.833	
Ninth day	1.009	...	0.656	0.433	

The percentage of carbon varies with the depth of sample. On the surface there is very little, and the bulk is found toward the centre of the casting, as shown by the following analyses:

Place of sample.	Combined C.	Temper C.	
Outside layer	Tr.	0.00	
Second layer	0.51	0.00	Ibid.,
Third layer	0.90	0.38	I, 1897
Centre of casting	1.40	2.38	p. 154.
Average of whole	0.74	1.56	

The microscope shows the outer layer to consist of crystals of ferrite, surrounded by filaments of silica and oxidized iron. The second layer shows ferrite and pearlite. The third consists of pearlite and cementite, while the centre contains cementite and separated carbon.

Specifications.—The standard specifications for mallea-bleized castings, as proposed by the American Society for Testing Materials, is as follows:

Sulphur not over 0.06 per cent.
Phosphorus not over 0 225 per cent.
Tensile strength not under 42,000 pounds per square inch.
Transverse strength not under 3,000 pounds per square inch on a 12-inch bar.
Elongation 2½ per cent. in 2 inches.
Deflection not less than ½ inch

Such castings have more resilience than steel, and also resist shock as well, but have less tensile strength. It is said that the open hearth gives 2000 pounds more tensile strength than the air furnace, but the air furnace can make metal of 52,000 tensile strength. The usual limits of tensile strength are 40,000 to 63,000.

Am Soc.
Test Mat.,
III, p. 204.

Action of Metalloids During Annealing.—The presence of **sulphur** delays annealing, since it tends to keep the carbon in the combined state and consequently opposes its elimination. It should, therefore, not greatly exceed 0 05 per cent. There is a tendency also to take up sulphur during annealing. **Phosphorus** tends to make the metal hard but fluid. Its limit is usually set at 0.2 per cent. **Manganese** is a hardener also, but it tends to eliminate sulphur during melting and so may become beneficial. It should never exceed 1.5 per cent.

Silicon should be kept low, in order that the castings may be perfectly white before annealing, but it should not be entirely absent, as some seems to be necessary in order to facilitate the separation of the temper carbon. The silicon which is required to release the carbon is about as follows:

Eng'r News,
XLIX,
p 531.

Heavy castings .... 0.45 per cent. Si
Ordinary castings . .. . 0 65 per cent. Si
Agricultural castings. .... .. 0 80 per cent. Si
Very light castings .. .. . .1 25 per cent. Si.

A composition which would show perfectly white fracture in small sections might show graphite in a large section which cools more slowly. Usually about 35 per cent. of the silicon in the pig is lost during melting, so the pig should contain 0.75 to 1.50 per cent., according to the class of work. Generally an additional 20 to 40 per cent. is lost during annealing.

Under ordinary conditions there is no considerable elimination of **carbon** during melting Any drop in percentage is due chiefly to the additions of less carburized metal in the form of steel, scrap and rabbles. Any attempt to eliminate the carbon in the furnace will result in excessive oxidation of the metal. The drop in carbon usually ranges from 0.2 to 0.4 per cent. The lower the carbon, the less annealing is necessary, but a good percentage is required to give fluidity and clear castings. The iron used may contain the usual content of carbon, *i e.,* from 2.75 to 4.00 per cent. The carbon must be wholly in the combined state in the castings, as graphite is not affected by annealing and its presence interferes with both strength and malleability.

Shrinkage.—The contraction of white iron is high, and the castings usually show a shrinkage of very near ¼ inch per foot, which is about double that of gray iron. During annealing, how- Iron Trade Review, 1900, Apr. 5. ever, the separation of solid carbon causes an expansion which more than compensates for the excess of contraction. As a result the net shrinkage of malleable castings averages somewhat less than that of gray castings.

Effects of Heat Alone.—Cold white iron may be considered as consisting of a mass of cementite and pearlite. When it is heated to about 1300 degrees F., which is a critical point, known as the "recalescence point," the pearlite becomes a homogeneous carbide, approximating the formula, $Fe_{24}C$, which contains 0.85 per cent. C. As the temperature rises above 1800 degrees F., the Fe_3C is decomposed and unites with the $Fe_{24}C$ to form a combination having a different degree of concentration corresponding to the formula $Fe_{14}C$, and containing 1.5 per cent C. Any excess of carbon will then separate as temper carbon, and give the iron a gray or black appearance sometimes known as "black heart" castings. This principle serves as the basis for the manufacture of "converted gray castings."

Converted Gray Castings.—Converted gray castings stand between gray iron castings and malleableized castings in that white castings are rendered somewhat malleable through annealing, but without any removal of carbon. Gray castings are made by heating white castings under a protecting cover to prevent scaling, to 1850 degrees F. for a few hours. The white surface of the metal becomes uniformly gray throughout, and the structure becomes

more dense   The metal is not so malleable as malleableized castings, but it can be forged and has a tensile strength of over
40,000 pounds.   The change which takes place is illustrated by
the following analysis :

Eng'r News,
XLIX, p 531

	C. C.	Temper C.	Si.	Mn.	P	S.
Before annealing............	2.00	0.72	0 71	0 110	0.039	0 045
After annealing............	0 82	2.75	0.73	0.108	0.039	0.040

Jour Frank.
Inst.,
1800, II,
p 227.

The presence of silicon assists in the conversion.   The higher
the silicon the less heat and time are necessary for the change
and the more thoroughly is it accomplished.   When Si is below
½ per cent., the change becomes very difficult.   The limit of
silicon is indicated by the appearance of graphite in the casting.

Effects of Various Methods.—The following table illustrates the effect upon white iron of various conditions of annealing :

Cover.	C. C.	Temper C.	Tensile strength.	Elongation.	Reduction.	Angle of bend, deg. F
White iron...........	3 69	0.19	15,600	0	0	0
Charcoal*	0.94	2 80	28,000	0	0	0
Cast borings*........	0 99	2 73	28,200	0	0	0
Wrought drillings*....	0.81	2 30	44,400	0	0	10
Lime†	0 64	1.79	38,600	1.90	3.15	75
Sand†	0.63	1 75	36,800	2.10	2.10	75
Boneash†	0.68	1 60	40,200	2.20	3 05	75
Hematite	0.73	0 36	32,600	4.00	2.10	80
Limestone	0 58	0.49	30,800	5 60	3.10	90
Ore, three times......	Tr.	0	33,20	10.80	7.90	180

* Luted.   † With access of air.

Decarburization appears to be most rapid and most complete in
an atmosphere of CO_2.

Tr. A S. C. E.
No 34, p

Charcoal pig in some proportion was long considered indispensable to the production of malleable castings, but an average
of 45 tests each, of coke and charcoal irons show that coke iron
gives castings which give higher tensile strength as well as better
elongation and reduction.

APPENDIX I.

SOME PRINCIPLES OF CHEMISTRY AND PHYSICS.
CHANGES IN MATTER.

Matter is a general term used by scientists to include all material substance which exists in the universe. It occurs in infinite varieties of form and nature, and the study of it forms the basis of such natural sciences as Chemistry and Physics. The quantity of matter in the universe is constant. It does not lie within the power of the scientist to either create or destroy matter. His power is limited to merely changing its form or condition, and it is through this power that all matter can be adapted to his purpose. Changes in matter may be classed under two general heads. When the change is superficial and does not alter the essential composition of the substance, it is denoted as **physical** when it is so fundamental as to alter the composition of the substance it is known as **chemical.** This distinction, when once understood is usually quite obvious, but in certain phases of the study it is difficult to tell where one ends and the other begins.

CHEMICAL CHANGES IN MATTER.

The science of Chemistry deals with the composition of matter. Any change in the composition of matter is known as chemical change, which must be carefully distinguished from mere physical change. Change in position, shape or temperature, etc., does not alter the nature of the substance and is therefore simply physical. For example, the cutting of timber, the forging of steel and the melting of lead leave each kind of material unchanged in its essential nature. There has been no change in the constitution of the substances involved. If, on the other hand, the wood be burned to invisible gases, or the metal be oxidized to brittle scale, there has been essential change in the nature of the substances. The old substances have passed into new combinations with other associations and properties, and there has been chemical change. Every chemical change is accompanied by radical change in the nature of the substance.

827

NATURE OF MATTER

All matter, however complex, must be made up of elementary substances By separation or **analysis** a chemist may discover what elementary substances combine to form a given mass of matter. He can go a step farther and prove the accuracy of his observations by **synthesis,** that is by causing similar elementary substances to combine, thereby reproducing the original kind of matter.

ELEMENTARY SUBSTANCES.

In separating the component parts of matter the chemist finds that certain components appear incapable of further separation. For example, oxides of iron can be easily separated into Fe and O, but neither can be made to undergo further division. These indivisible components are assumed by scientists to be reduced to their lowest terms, and are therefore called **elements.** It is not beyond the bounds of possibility that some or all of what are now considered elements may some time be broken up still further, but so far they have resisted all the skill of science, and so may be considered by us in all respects as strictly elementary bodies

Atoms.—While an element cannot be separated into component elements, it may be capable of very fine subdivision. A bar of iron, for instance, may be reduced to iron filings, and each grain of filings may be divided still further. This subdivision may be conceived as continued until the most minute particle which can exist is reached. This ultimate particle of an element is called an **atom.** Atoms of all kinds seem to be made up of still more minute particles called **corpuscles,** but for the present discussion an atom may be considered the final product of subdivision.

Atomic Weights.—Such a tiny particle of matter as a chemical atom could hardly be considered as having much size and weight, yet it must necessarily have both. It is impossible, of course, for us actually to weigh in pounds and ounces such imponderable masses, and yet we have a list of the atomic weights of the elements These weights are not absolute, but relative, and represent the comparative weights of the simplest combining proportion of each element. Scientists have concluded that the smallest weight in which a given element has ever been found in combination must be the weight of its smallest subdivision, that is, its atom. In default of a scale of weights sufficiently delicate for such comparisons it was necessary to improvise one, the basis of which is the lightest known atom. Hydrogen is the element that has the distinction of being found in combination with other elements in the smallest proportion by weight, hence the weight of its atom is taken as the standard 'and designated as 1. The combining weights of all the other elements have been determined very carefully on this basis.

The combining weights of the elements must not be confused with absolute weight, or weight per cubic foot. The latter is known as the **specific gravity** of the substance, and is usually expressed as the ratio of the weight of a unit volume of the substance to the weight of the same volume of water. For example, a cubic foot of water weighs 62.4 pounds and a cubic foot of cast iron weighs 450 pounds. The specific gravity of cast iron, therefore, is $\frac{450}{62.4} = 7.2$.

Chemical Notation.—By common consent the notation of chemical elements has been abbreviated to symbols which are usually the initial letters of the English or Latin name. The use of the symbol alone signifies that only one atom of the element is involved. Two or more atoms are indicated by a small subscript figure written after the symbol. For example, O stands for oxygen and Fe stands for iron, (Latin, ferrum). Fe_2O_3 stands for a compound of two atoms of iron and three of oxygen The atoms of elementary substances appear to unite with each other, usually in pairs, to form molecules of great stability, as O_2, N_2, etc.

Nature of Elements.—The elements with which most of us are best acquainted are probably the metals. Most of the metals which we use daily are elements, except a few, such as bronze, brass, babbitt, solder, etc., which are alloys of two or more metals. Elements are not all metals, however, nor are they even solids at ordinary temperatures. Some exist as liquids and several as gases.

Mixture of Elements.—Elements may usually be mixed to-

gether under ordinary conditions without undergoing change. A
conspicuous example of this is atmospheric air which we breathe.
Air is a mixture of the two gases, oxygen and nitrogen, in the
proportion of 20.9 parts to 79.1 parts respectively by volume, and
23.3 parts to 76.7 parts by weight. There are traces of other
gases in the air, but the total is small in comparison to the quantities of O and N. Although air is simply a mixture of O and N,
these proportions hold good the world over. The components of
air may be separated and examined. They are found to be perfect gases under all ordinary conditions of temperature and pressure and respond to all of the laws to which gases are subject.
Air, likewise, is a perfect gas, and is governed by all the laws of
gases. Hence we see that these two gases, when put together,
form another perfect gas, which, since it partakes of the properties of each must be simply a mixture.

COMPOUNDS.

Chemical elements enter into definite combination with each
other in proportion to their atomic weights. These combinations
are known as **compounds.** In forming a compound the combining elements involved lose their identity completely, and form a
new substance that may differ widely from them. For instance,
two gases may unite to form a liquid, or a gas and a liquid may
unite to form a solid. It is such radical changes in characteristics
that enable us to determine whether a given change is chemical
or merely physical.

Molecules.—The smallest particle of a compound which can
exist, that is, one which cannot be further subdivided without
separating it into its component atoms, is called a **molecule.**
The molecule bears the same relation to the compound that the
atom does to the element, in that it is the smallest conceivable
portion of it. Each molecule of a given compound, however, is
composed of the same number of the same kind of atoms as every
other molecule.

Quantivalence.—Atoms have the property of uniting with
other atoms according to a certain numerical law, which is designated as **quantivalence.** For example, the hydrogen atom appears
to possess but a single bond for uniting with other elements, hence

it is denominated a univalent element Oxygen possesses two bonds and is said to be bivalent, nitrogen three, and is trivalent, silicon four and is quadrivalent, phosphorus five and is quinquivalent, chromium six and is hexivalent. Under certain conditions two adjacent bonds in a given atom may unite with each other and, in consequence, a hexivalent atom may appear temporarily to be quadrivalent or bivalent, and a quinquivalent element may appear trivalent or univalent Such an element may be described as unsatisfied or unsaturated, and, in consequence, usually forms additional relations readily

Composition of Water.—One of the simplest compounds, and one that is very familiar to all, is water. Like air, it is composed of two gases, but, unlike air, the gases are combined and not merely mixed. The components of water are hydrogen and oxygen. Since the hydrogen atom possesses only one bond and the oxygen atom has two, the law of quantivalence makes it necessary that the water molecule shall consist of two atoms of hydrogen and one of oxygen, or some simple multiple It is generally considered that the simplest formula is the true one, hence it is written H_2O, in accordance with the chemical notation, which shows at a glance the chemical composition of substances.

Classification of Compounds.—H_2O may be considered as a type of compounds in general. The hydrogen may be replaced by any of the simple elements except Fluorine, and by many combinations made up of two or more elements. Such a group replacing hydrogen in any of its compounds is known as a **basic radical.** On the other hand, the oxygen of H_2O may be replaced by various elements or groups of elements, and such replacing groups are known as **acid radicals.** Acid radicals, replacing the oxygen of water, give a class of compounds known as **acids.** When acid and basic radicals unite, the combinations are known as **salts.** The replacement of the hydrogen of water by other elements brings into existence a class of compounds known as **oxides.** When the replacing elements are simple metals, the metallic oxides so formed are known as **bases.**

OXIDATION AND REDUCTION.

The process of attaching oxygen to an element is known as **oxidation,** and the element is said to be **oxidized.** The removal of oxygen from an oxidized compound is called **reduction,** and the operation is described as "complete" or "partial," according to whether the oxygen is wholly or only partially removed The oxidation of a substance is a chemical operation and the product of the action is always very different from the initial components. For example, the oxidation of iron produces rust or scale, which presents no resemblance to either a strong, tough metal or a gas. The combustion of wood or coal is nothing else than the phenomenon produced by the uniting of the oxygen of the air with the carbon and hydrogen of the fuel.

Oxidizing Agents.—Oxygen is a particularly aggressive element, and as it is present everywhere in the atmosphere, oxidation of other elements frequently takes place spontaneously. The rusting of iron is the result of slow, spontaneous union going on between the metal and the atmospheric oxygen. Any protective coating that keeps iron from contact with the oxygen of the air' will prevent rusting Oxidation of substances in contact with air would be much more rapid if the aggressiveness of oxygen were not restrained by the presence of nitrogen. Nitrogen is as inert as oxygen is active, and, as it comprises four fifths of the atmosphere, it dilutes its aggressive companion to such an extent that its activity is enormously reduced. For example, we know that oxidation of iron by the atmosphere, in the familiar form of rusting is a comparatively slow process. If, however, there be prepared a jar of pure oxygen, free from the restraining influence of nitrogen, and a piece of iron wire, heated to redness at one end, be thrust into it, the wire will burn as actively as does a splinter of wood in ordinary air. It will give off a bright, sparkling light, and in a few seconds be changed to bits of black scale, closely resembling the scale from a rolling mill.

We have seen that oxidation of elements may take place under certain conditions through direct contact with oxygen gas It may also take place by a method of exchange between oxidized and unoxidized substances. A necessary condition to the reaction, however, is that under the existing circumstances the unoxidized substance must have a stronger attraction for the oxygen than does the oxidized substance For example, at certain moderately high temperatures, carbon dioxide, CO_2, will give up part of its oxygen to metallic iron. The exchange may be represented graphically by this equation,

$$CO_2 + Fe = CO + FeO,$$

in which we see that one of the atoms of oxygen has left the CO_2, and attached itself to the Fe, thereby partially oxidizing the iron to FeO, and partially reducing the carbon to the condition of CO. As a matter of fact, this reaction is not so simple in reality as the above equation would indicate. In this simplified form it serves only as an illustration of the general principle involved. It should be observed that in this case of oxidation of iron by transfer of oxygen, there is of necessity a corresponding reduction of the carbon. In the case of oxidation by direct contact with oxygen gas there was no accompanying reduction, because the oxygen used was already free. In oxidation by means of oxidized substances, it is evident that reduction must precede oxidation in order that the necessary oxygen may be available for the purpose. Since both reactions take place, it is customary to designate the operation according to the reaction that we are endeavoring to accomplish.

HEAT OF COMBUSTION.

In the process of combining elements to form chemical compounds, heat is given off by some reactions and absorbed by others The quantity of heat has been measured many times, and it is found that a given quantity of material entering into a given reaction always involves the same quantity of heat. Those reactions that give off heat are called **exothermic** reactions, and those that absorb heat are called **endothermic** reactions. The union of oxygen with the other elements with which we have to deal is always exothermic. It is this property of oxidation that makes fuels efficient heat producers. Fuels, such as wood, coal, petroleum, etc., are made up of carbon and hydrogen. The process of combustion is merely the rapid oxidation of these two elements by the oxygen of the atmosphere. Such oxidation does not take place at ordinary temperatures. There is a definite temperature for each combustible substance known as the "ignition point," at which rapid oxidation begins. The combustion then continues as long as the temperature of the flame keeps above the ignition point of that substance. Spontaneous combustion occurs when internal heating raises the temperature of a combustible to the ignition point. Decay of organic substance is a slow oxidation at ordinary temperatures. The amount of heat given out, however, is the same as if the substance had been burned by rapid oxidation.

COMBINATIONS OF OXIDES.

As was observed above, the oxides of metals are usually termed bases. The chief bases with which we will deal are the oxides of iron, manganese, calcium, magnesium, sodium and potassium. The non-metallic elements are usually acid-makers, such as silicon, sulphur, phosphorus, carbon and nitrogen. These acid and basic radicals when in solution react upon each other and produce a class of compounds called salts. For example, CaO may be considered as uniting with SO_3 to form the salt, calcium sulphate, $CaSO_4$. In the same way, we may consider that MgO and CO_2 unite to form magnesic carbonate, $MgCO_3$, or Na_2O and SiO_2 to form sodic silicate, Na_2SiO_3. When these oxides are brought into contact in a state of fusion, they show the same affinities as when in solution, although the outward effect is vastly different. The slags from metallurgical operations are the result of such fusion.

Slag is a general term which covers a various and complex class of bodies. The name is applied to the scoria which is present in all metallic fusions and which may vary in composition from fused metallic oxides, carrying little or no acid constituents, to highly siliceous compounds having but a small percentage of bases. The slag from an iron blast furnace is usually composed largely of calcium silicate, although the calcium may be partially replaced by magnesium and the silica by aluminum.

REDUCING AGENTS.

We have seen that oxidation may take place with or without a corresponding reduction. On the other hand, reduction is always accompanied by a corresponding oxidation. The reduction of iron oxide, for example, may be accomplished by means of several reducing agents, but in each case, the latter undergoes corresponding oxidation, thus,

$$Fe_2O_3 + 3H_2 = Fe_2 + 3H_2O$$
$$Fe_2O_3 + 3C = Fe_2 + 3CO.$$
$$Fe_2O_3 + 3CO = Fe_2 + 3CO_2.$$

The occurrence of these interchanges demands as a rule, certain specific conditions. The ruling condition is usually the temperature, but it is also of first importance that the different substances should be brought into intimate contact for a sufficiently long period for the interchange to take place Each of the above reducing agents has a critical temperature below which it refuses to act. Hydrogen will attack the oxide at temperatures lower than C or CO. It is active at all temperatures above the boiling point of water, 212 degrees F. Carbon monoxide does not begin to reduce iron until about 400 degrees F. is reached, while carbon needs a temperature of nearly 800 degrees F. At temperatures above redness, the action of CO may be reversed and the reduced iron will be oxidized by the CO_2 formed by the reaction. As a matter of fact, most chemical reactions are reversible under other conditions of temperature and pressure. The point of equilibrium under a given set of conditions depends upon the relative concentrations of the agents present, the nature of the reaction being determined by the agent having the greatest concentration. This fact is the basis for the **Law of Mass Action.** Reactions that produce compounds which neutralize the reagents will check themselves, unless the neutralizing compounds are continually removed from the point of action. It is evident, therefore, that reactions cannot be complete unless they produce insoluble or volatile compounds.

DESCRIPTION OF CERTAIN ELEMENTS.

Only a few of the seventy odd elements that have been distinguished are involved in the metallurgy of iron It is inadvisable for a beginner in the study to burden his mind with other elements than those which are essential A brief description of the most important facts concerning the few that are necessary is given below.

OXYGEN. O.

Oxygen is a transparent, colorless gas with neither taste nor smell. In appearance it cannot be distinguished from ordinary air It is the most abundant substance in the universe. It comprises eight-ninths of all the water, one half of all the land and one fifth of all the air, besides entering largely into the composition of plants and animals. It combines with all other elements except one, which indicates that it is the most active or most popular of all the elements. In the act of combining with other elements, it gives out heat and frequently light, producing the phenomenon known as combustion. Its activity is such that if the oxygen of the air were not diluted with four times as much nitrogen, no substances of an oxidizable nature could long withstand its presence, and we would be constantly subjected to spontaneous combustion.

Oxide Compounds.—When oxygen enters into combination with the other elements singly, the resulting compounds are known as oxides. Many such oxides, when combined with water, become members of the class of compounds known as acids. For example,

$$SO_3 + H_2O = H_2SO_4, \text{ sulphuric acid.}$$
$$CO_2 + H_2O = H_2CO_3, \text{ carbonic acid.}$$
$$SiO_2 + H_2O = H_2SiO_3, \text{ silicic acid.}$$

In fact the word oxygen means "acid-maker." The above examples show that some acids at least are but hydrated oxides. Oxides that can be obtained by dehydrating an acid in this way are known as anhydrous oxides or anhydrides. For example, sulphuric trioxide, SO_3, is called sulphuric anhydride, and carbonic oxide, CO_2, and silicic dioxide, SiO_2, are sometimes called carbonic and silicic anhydrides.

The great duty of oxygen in the universe is to support combustion. Oxygen, itself, does not burn, that is to say it cannot unite with itself. But it is absolutely necessary to the combustion of other materials, since combustion is but the oxidation of matter.

NITROGEN. N.

Like oxygen, nitrogen is a transparent, colorless, tasteless, odorless gas not easily to be distinguished from air. It is not surprising that both oxygen and nitrogen resemble air, since it is but a simple mixture of the two. In its behavior, nitrogen is very unlike oxygen. While oxygen is the most active element, nitrogen is very inert. It is singularly indifferent to forming compounds with other elements. It does form such compounds, but only with reluctance. This is well illustrated during the passage of the blast of air through the blast furnace. During that time the oxygen of the air is entirely consumed, losing its identity entirely, but the great bulk of the nitrogen passes out of the furnace unchanged. A small portion, however, unites with carbon at the high temperatures of the furnace and forms cyanogen, CN, which is the basis of all cyanides. ·

Although when mixed with oxygen nitrogen forms air, which supports combustion and life, yet when alone it extinguishes flame, purely for lack of oxygen. It does not, therefore, play an important part in metallurgical operations, although its universal presence makes it necessary that it be fully understood and allowed for.

HYDROGEN. H.

Hydrogen, when pure, is a transparent, colorless, tasteless, odorless gas, not easily to be distinguished in appearance from oxygen, nitrogen or air. Like nitrogen, it is incapable of supporting combustion. The flame of what are ordinarily considered combustibles will not live in an atmosphere of hydrogen. In the presence of oxygen, however, it is itself exceedingly inflammable, that is to say, it has a very strong affinity for oxygen. The ignition point, or the point at which the two begin to combine, is comparatively low, and the heat of the reaction is very high, the reaction giving off more heat than any other known. Burning hydrogen is often used for producing very high temperatures. The product of such combustion is hydrogen's most important compound, water, in the form of very highly heated steam, thus,

$$2H_2 + O_2 = 2H_2O.$$

A mixture of hydrogen with oxygen or air is very explosive. Its affinity for oxygen, however, is not confined to oxygen in the free state, but it has a strong attraction for oxygen when combined with other elements. It is therefore a powerful reducing agent, one of the most powerful known. It attacks oxides of iron, and reduces them to the metallic condition with ease at comparatively low temperatures with the accompanying formation of water. It is the lightest substance known, and, as has been stated, the weight of its atom serves as the standard from which we compute the relative weights of the atoms of the other elements.

CARBON. C.

Carbon is a very abundant element and one that is extremely important in metallurgical operations. It is widely distributed and appears under many guises. For example, charcoal, graphite or plumbago, coke, lampblack, soot and even the diamond, are but different modifications of this same substance. When combined with other elements, in the form of certain hydrocarbon gases, it forms also coal, wood and all other organic substances. It is even a constituent of certain rocks, such as limestone, dolomite and spathic iron ore. Since it is the chief constituent of wood, coal, coke, charcoal and most combustible gases, it is plainly the basis of all fuels, and is, therefore, our chief heat-producing agent.

Varieties of Carbon.—Carbon occurs in three distinct states which do not bear much resemblance to each other, and which are known as allotropic modifications.

In a highly crystalline form it appears as the **diamond,** where it is the hardest substance known. It is clear, transparent and highly refractive to light rays. It is not affected by ordinarily high temperatures when out of contact with oxygen. In the electric arc, it softens, swells and forms black coke. When heated in the presence of oxygen it burns to CO_2.

As a semi-crystalline substance, carbon appears in the form of **graphite, plumbago** or **black-lead.** Graphite is used to make lead pencils and "black-lead" crucibles for melting metals. It is largely in the form of graphite that carbon appears in pig iron.

The **amorphous** or **non-crystalline** variety of carbon is the form that most interests the metallurgist. This variety includes hard coal, coke and charcoal. The soft coals also contain from 50 to 75 per cent. of carbon, and from 20 to 40 per cent. of combustible gases. These gases also contain a large percentage of carbon combined with hydrogen, and are generally known as hydrocarbon gases. The chief of these hydrocarbon gases are,

Methane, or marsh gas, having the formula, CH_4, and Ethylene, or olefiant gas, having the formula, C_2H_4.

Oxides of Carbon.—Carbon has a strong affinity·for oxygen at elevated temperatures. When burned in the presence of air or free oxygen it is capable of forming two oxides, CO, known as carbonous oxide or carbon monoxide, and CO_2, known as carbonic acid, carbon dioxide or carbonic anhydride. Both CO and CO_2 are fixed gases, transparent and invisible, but they differ radically in character. CO is an unsaturated compound, capable of taking up another atom of oxygen and forming CO_2. In other words, it is combustible. It burns with a blue flame which is familiar to us in hard coal or coke fires, giving very little light but much heat. It is extremely poisonous when inhaled and is responsible for the many deaths through asphyxiation CO_2, on the other hand, is a completely saturated compound and can take up no further additions of oxygen. It is therefore incombustible.– It is a heavy gas, being about $1\frac{1}{2}$ as heavy as air. It has a tendency to collect in depressions in the earth, such as caves, wells, enclosed valleys, etc., and while it is not poisonous, it is a source of danger to men or animals entering such depressions. Continued stay in its presence will result fatally, as it will not support life or combustion.

Combustion of Carbon.—Carbon, in order to burn to the condition of CO_2, must have air equal to or in excess of the requirements indicated in the equation,

$$C + O_2 = CO_2.$$

Free access of air always gives CO_2 as the product of burning carbon. In forming CO_2, carbon gives up all of the heat of which it is capable. It has been determined that pure carbon, burning to CO_2 gives approximately 14,550 British Thermal Units, that is, one pound of carbon burned to CO_2 will raise 14,550 pounds of water 1 degree F. in temperature. When carbon is burned with an insufficient supply of oxygen some CO is formed, the amount depending upon the degree of insufficiency. When there is present just half the amount of oxygen that is needed to form CO_2, the whole of the product of combustion of carbon will be CO, thus,

$$2C + O_2 = 2CO.$$

It has been determined that carbon, burning to·the condition of

CO, develops 4450 B. T. U., which is only about 30 per cent. of the possible total when the combustion is complete. Incomplete combustion of fuel, therefore, is evidently very uneconomical. CO, however, is a combustible gas, and is capable of taking up another atom of oxygen. In doing so, each pound of carbon gives out 10,100 B. T. U., which is the other 70 per cent not developed during the formation of CO. It appears, therefore, that owing to the absorption of the latent heat of gasification the addition of one atom of oxygen develops far more heat in one case than in the other. This unequal development of heat has certain advantages If a gaseous fuel is desired, it may pay to sacrifice 30 per cent. of the heat in coal in order to develop a combustible gas which still holds in a potential state 70 per cent. of the original heat of the coal. This idea forms the basis of the widespread practice of converting coals into producer gases for combustion in gas-burning furnaces.

Carbon Transfer.—Carbon monoxide may be produced from carbon without the intervention of air or free oxygen. When CO_2 is brought into contact with carbon which is heated to redness, a portion of the carbon is dissolved by the second atom of oxygen which forms the CO_2, and CO is the result, as indicated in the following reaction,

$$CO_2 + C = 2CO.$$

This is a refrigerating or endothermic action, owing to the fact that 10,100 B. T. U. is absorbed in separating the oxygen from the CO_2, and only 4450 B. T. U. is liberated when it combines with the fresh carbon. Such a reaction cannot continue long unless heat be supplied from some external source.

Carbon as Reducing Agent.—As a reducing agent in metallurgical operations, carbon has a value second only to its worth as a fuel. Its power to act simultaneously as a producer of heat and as a reducing agent makes it a prime requisite to the smelter. At temperatures above 700 degrees F., solid carbon has the power of extracting oxygen from oxides of iron, thus,

$$Fe_2O_3 + 3C = Fe_2 + 3CO,$$

but as the reduction of iron ore is pretty well completed by the time it reaches that temperature, it follows that the above reaction is not very important to the iron smelter. Nevertheless, carbon is the chief reducing agent in the manufacture of pig iron, but it is in the form of CO that it is most active. CO is capable of taking up oxygen from some ores of iron at temperatures as low as 400 degrees F. It is the power to act at such low temperatures that enables it to have the reduction nearly done before the solid carbon can act. The action of CO upon oxide of iron may be represented thus,

$$Fe_2O_3 + 3CO = Fe_2 + 3CO_2,$$

although the action is not as complete as this equation would indicate. The heat evolved by this reaction more than balances that absorbed, and hence the reaction is exothermic.

IRON. FE.

Iron is an element that is familiar to all. It is naturally a soft, tough, fibrous metal when pure, as in wrought iron or soft steel. When alloyed with small quantities of carbon, however, it becomes harder and less tough, as when it appears in the form of steel. When it is alloyed with 3 to 4 per cent. of carbon and almost as much silicon, together with other elements, such as phosphorus, sulphur and manganese, we have the familiar form of cast iron, which differs very widely from the natural metal.

Oxides of Iron.—Like other metals, iron readily forms oxides and salts. It is very susceptible to the attack of oxygen, and forms oxides even at ordinary temperatures. The oxide is commonly called rust. The usual red iron rust is the **sesquioxide of iron,** Fe_2O_3, usually called ferric oxide. This compound in the form of ore is the source of about 80 per cent of the iron produced in the world. Besides the ferric oxide, a **ferrous oxide,** FeO, is found in combination, although it cannot be isolated. It is in this form that iron is generally present in blast furnace slags. Iron forms also a **ferro-ferric oxide,** Fe_3O_4, which is magnetic and is practically identical with roll scale and magnetic iron ore.

SILICON. SI.

Next to oxygen, silicon is the most abundant element in the earth. Nevertheless, comparatively few persons are familiar with its appearance. This is due to the fact that it clings very tenaciously to its compounds and is rarely isolated. It is said to occur in three modifications, like carbon. The amorphous variety is a brown powder which burns readily in air.

Silica.—Silicon forms one oxide, silicon anhydride, or silica, having the formula SiO_2. The attraction between silicon and oxygen is very strong, and much heat is given out during the combustion. Pure silica is a white, earthy substance and occurs in nature as quartz, rock-crystal and fine, white sand When colored by iron or other metals it appears as flint, jasper, agate and some of the precious and ornamental stones. Combined with various bases it is the chief component of all igneous rocks and also of all manufactured and natural glass. It is usually the chief acid factor of metallurgical slags, and the chief ingredient of refractory firebricks.

Silicon in Pig Iron.—During the smelting of iron, various quantities of silicon are reduced from the silica and alloy with the iron. The effects of this silicon upon the quality of pig iron have an enormous bearing upon the metallurgic art.

ALUMINUM. AL.

Aluminum is in some ways related to silicon. While it is not as abundant as silicon, yet they are generally associated in nature, and so occur together in artificial products that are obtained from earthy materials. In this way aluminum in its oxidized form enters largely into the composition of slags. Aluminum is a metal, and its oxide, Al_2O_3 is generally considered a base and may unite with silica, forming silicates of aluminum, such as clay Yet when there is a plenitude of other bases and a scarcity of silica, alumnic oxide appears to have the faculty of acting as an acid, as it assists the silica in assimilating the bases.

Alumina.—Aluminum forms one oxide, Al_2O_3, which is called alumina. It is a hard, granular substance much like silica, and occurs in nature as corundum, ruby, emerald, etc. It occurs also in most igneous rocks and is the essential ingredient of all clays.

CALCIUM. CA.

Calcium is a yellowish-white metal, which owing to its strong affinity for oxygen is difficult to reduce and is rarely seen in its elemental form. Its properties are not such as to make it especially useful, and therefore there is no incentive to separate the metal. It has its greatest usefulness when in the form of its oxide.

Lime.—Calcium is bivalent and forms one oxide, CaO. This compound is the familiar substance which is called lime, and is used largely in agricultural and building pursuits. It is made by heating limestone, $CaCO_3$, to high temperatures. The CO_2 is liberated by the heat, and the CaO is left behind in white, friable lumps, which is known as **burntlime** or **quicklime.** When this is exposed to the air for considerable periods it absorbs moisture and falls to powder. It is then known as **slacked lime.** A clear solution of lime in water is used as medicine under the name of **limewater.** When mixed with quantities of water insufficient for complete solution, it is known as **milk of lime.** When limestone is used as a furnace flux, the CaO is the base which unites with the siliceous gangue to form the slag.

MAGNESIUM. MG.

Magnesium is a white, lustrous metal, which burns readily in air with a dazzling white light. It is the basis of all flashlight preparations for photographic work. Magnesium and its compounds resemble closely those of calcium. They both belong to a group of elements known as the alkaline earths. Magnesium occurs in nature in large quantities as **magnesite,** $MgCO_3$, and as **dolomite,** $Mg(Ca)CO_3$. Both substances when calcined are largely used as refractory materials for lining steel-making furnaces. In the form of dolomite magnesium enters the blast furnace as flux, where in the form of MgO it replaces some of the CaO in the slag.

Magnesia.—Magnesium forms one oxide called magnesic oxide, or magnesia, MgO, which is metallurgically its most important compound. It is a white solid and is the product alike of the combustion of flashlight powder and of the calcination of magnesite.

MANGANESE. MN.

Manganese is a metal of grayish-white color, hard and brittle. It alloys readily with iron, and is generally produced as such **an** alloy. Its uses are mainly confined to its alloys, as the pure metal does not possess the usual qualities desirable in a metal. Its most used alloys are spiegeleisen and ferromanganese, which are universally employed in steel manufacture.

Oxides of Manganese.—Manganese forms several oxides. MnO_2, manganese dioxide, is its chief ore. MnO, manganic oxide, is the form in which it enters into combination with silica in slags.

THE ALKALIS. NA AND K.

Sodium.—Sodium is a soft, silvery white metal, which cannot exist either in air or in water. Its most usual compound is the chloride which is our common salt. Its hydrate is largely used in making soap and its silicates in making glass.

Potassium.—Closely related to and resembling sodium is the metal potassium. They frequently occur together and their compounds are used for similar purposes, hence they are often classed together as the alkalis. They both occur in small quantities in the ash of coal and wood and sometimes in the gangue of ores and limestone. Thus they find their way into blast furnace slags in small quantities. The alkalis show an especial affinity for cyanogen, and the resulting alkaline cyanides are supposed to have considerable effect upon the reactions in the furnace.

PHOSPHORUS. P.

Phosphorus, when pure, is a colorless, transparent, wax-like substance. When heated gently out of contact with the air, it changes to an allotropic condition known as red phosphorus. It unites readily with oxygen, being very inflammable, and hence finds wide use in the manufacture of matches. It occurs in nature chiefly as calcic phosphate, and as such is largely used as a fertilizer. The universal presence of phosphates in all the materials which go to make up a furnace charge makes phosphorus an important element from a metallurgical standpoint.

Oxides of Phosphorus.—Phosphorus forms several oxides according to the degree of oxidation. The chief oxides are phosphoric anhydride, P_2O_5, sometimes called the pentoxide, and phosphorus anhydride, P_2O_3, known also as the trioxide. The oxides of phosphorus are acid radicals, and demand basic associates, yet their affinity for bases is not so strong as that of silica, and therefore they are readily displaced by silica in slags. The displaced oxides are readily reduced and the phosphorus is found in the iron in consequence. Very basic slags are capable of holding phosphorus, but blast furnace slags are rarely sufficiently basic to do so.

SULPHUR. S.

Sulphur at ordinary temperatures is a brittle solid of light yellow color, but at higher temperatures takes on allotropic modifications. It burns freely in air but is less inflammable than phosphorus. It forms two oxides, SO_2, sulphuric dioxide and SO_3, sulphuric trioxide or anhydride. Both substances are volatile at ordinary temperatures, but are quite irrespirable, and are largely used for killing disease germs by fumigation. Sulphur is the basis of sulphuric acid, and a very large class of compounds known as sulphates

Sulphur occurs in ores usually in the form of iron pyrites, FeS_2, a brassy looking mineral, showing bright, crystalline facets. It always acts as an acid radical whether in the elemental or oxidized condition. The element as well as its oxides is volatile. All of these conditions have a bearing on the action of sulphur in the hearth of a blast furnace.

Since elemental sulphur is an acid radical, it may or may not enter the iron. Unlike the other elements it may unite directly with calcium, forming CaS, and enter the slag. Its oxides are always found in the slag. The quantity of sulphur that may enter the slag varies with the quantity of lime present, hence the amount of lime used has an important influence upon the quantity of sulphur entering pig iron.

Since sulphur and its oxides are volatile at high temperatures, it follows that a very hot hearth will retain less sulphur in its pig iron than a cold one. These two factors, heat and lime, are the means for controlling the quantity of sulphur in pig iron.

CERTAIN ELEMENTS AND SOME OF THEIR PROPERTIES.

Substance.	Name.	Chemical symbol.	Atomic weight.	Specific gravity.	Quantivalence.	Oxides formed	Nature of oxide	Mol wt.	B. T. U. evolved in forming oxide per pound element.
Gas..	Hydrogen ...	H	1	...	1	H_2O	Colorless liq.	18	62,000
Metalloid...	Carbon	C	12	..	2	CO	Colorless gas	28	4,450
Metalloid. .	Carbon	C	12	..	4	CO_2	Colorless gas	44	14,550
Gas...... ..	Nitrogen....	N	14	..	*	*	*	*	*
Gas	Oxygen....	O	16	...	2	..	......	..	.
Metal	Sodium ...	Na	23	0.97	1	Na_2O	†	62	3,950
Metal........	Magnesium	Mg	24	1 74	2	MgO	White solid	40	10,755
Metal	Aluminum	Al	27	2 67	3	Al_2O_3	White solid.	102	13,086
Metalloid...	Silicon.....	Si	28	...	4	SiO_2	White solid	60	11,571
Metalloid ..	Phosphorus	P	31	..	5	P_2O_5	White solid.	112	10,605
Metalloid....	Sulphur	S	32	..	4	SO_3	Pungent gas	64	3,896
Metal......	Potassium ..	K	39	0.86	1	K_2O	†	94	2,266
Metal..... .	Calcium....	Ca	40	1 58	2	CaO	White solid	56	5,917
Metal.	Manganese	Mn	55	8 00	2	MnO	†	71	2,975
Metal........	Manganese	Mn	55	..	4	MnO_2	Black solid	87	4,100
Metal	Iron... .	Fe	56	7 90	2	FeO	†	72	2,111
Metal	Iron........	Fe	56	...	3	Fe_2O_3	Red solid	160	3,144
Metal........	Iron	Fe	56		2&4	Fe_3O_4	Gray-black	232	2,900

* Forms several oxides at different valencies.
† Occurs only in combination.

PARTIAL COMPOSITION OF EARTH'S CRUST BY WEIGHT

(Showing the Relative Proportion of Different Elements Present)

Calculated by
F. W. Clark,
Bulletin 78,
U. S. Geol. Sur.
pp. 34-43.

	Per cent.		Per cent
Oxygen	49.98	Potassium	2.23
Silicon	25.30	Hydrogen	0.94
Aluminum	7.26	Titanium	0.30
Iron	5.08	Carbon	0.21
Calcium	3.51	Phosphorus	0.09
Magnesium	2.50	Manganese	0.07
Sodium	2.28	Sulphur	0.04

PHYSICAL CHANGES OF MATTER.

From Kemp's
Ore Deposits.

While chemical change plays, as we have seen, a very important part in the metallurgy of iron, yet it is so interwoven with physical change that it would not be a simple matter to draw the line sharply between the respective utilities of the two. The study of blast furnace phenomena would be very incomplete if it did not include a study of some of the physical changes of matter, particularly those of air and some of the fixed gases which are so closely connected with the process of combustion

PHASES OF MATTER.

All matter is supposed to be capable of assuming the three conditions or phases, viz., solid, liquid and gaseous. Those substances which appear to us as solids at ordinary temperatures, may be made liquid or gaseous by a sufficiently elevated temperature. Generally liquids may be frozen into solids or converted into vapors, and gases may be condensed to liquid and solid forms. The state which we associate with familiar objects is, therefore, not necessarily the only one inherent in them, but merely the one in which they exist at ordinary temperatures Some substances are made to assume other states only by intense extremes of temperature. For example, iron at ordinary temperatures is solid. To make it liquid requires a temperature of more than 2500 degrees F., and to render it volatile requires a much higher temperature. On the other hand, some substances, such as water, for example, are capable of assuming all three states within a comparatively narrow range of temperature. At ordinary temperatures water is a liquid, at 32 degrees F. it is ice, and at 212 degrees F. it is turned into vapor. Such changes, it must be remembered, are merely physical and do not at all affect the composition of the substance.

THE PHASE LAW.

Every substance, whether in the state of solid, liquid or gas, is subject to three variables, whose effects are mutually interdependent, viz. temperature, pressure and volume. When the substance is present in only one state or phase, two of the variables must be fixed upon before the third can be determined. The third follows as a consequence of the two which are fixed. For example, a given weight of gas may be known to have a certain temperature, but its pressure and volume will still be indeterminate and subject to mutual variations. If, however, a given pressure is also fixed upon, the gas will necessarily occupy a certain definite volume, or if its volume can be fixed, its pressure can be determined. The gas is, therefore, said to have two **degrees of freedom,** and this statement is true of every uniform substance when present in only one phase.

With the addition of each phase, however, the number of degrees of freedom is reduced by one. When two phases are present the fixing of one variable determines the other two. For example, if water and its vapor exist in contact at atmospheric pressure, a definite volume of vapor will form at that temperature. If now, the pressure be diminished, for example, additional vapor can form and the temperature will fall and equilibrium be established again at larger volume and lower temperature. When ice, water, and water vapor are present, there are no degrees of freedom, as they can exist simultaneously only at one temperature, pressure and volume. The addition of other substances, however, adds a degree of freedom for each new component. With two components, such as water and alcohol the presence of two phases still permits two degrees of freedom, in accordance with the following **Phase Law:**

No. phases + no. deg freedom = no. components + 2, which is true for any number of components and phases. This is a natural consequence of Dalton's **Law of Partial Pressures,** which states that in any volume of mixed gases each gas acts independently of every other as if it filled the entire space alone, and the pressure it exerts is therefore in proportion to the quantity present. The sum of the partial pressures of the gases present is always equal to the whole pressure exerted by the mixture.

THE STATES OF MATTER.

As already stated, all matter, in whatever phase it occurs, is made up of minute particles styled "molecules." These molecules are held together by a mutual attraction which is usually designated as "cohesion." The strength of this force is always the same for any given substance under given conditions, but it varies widely for different substances and also for the same substances under different conditions. Indeed, the striking differences between the various phases of a given substance are due chiefly to difference in cohesion.

Solids.—The solid state of matter is always characteristic of its lower temperatures and appears to have the cohesive force more marked than that of the other phases. It is always characterized by comparative rigidity and tenacity and is not readily changed in shape without fracture. This condition is attributed to the pressure of a rigid, inflexible state of the cohesion, which does not permit of distortion without rupture. Yet the molecules are supposed to be separated from each other and are capable of free motion within a certain radius.

Liquids.—The liquid state of matter usually follows when the solid phase is subjected to a sufficiently high temperature. It is characterized by greater mobility, enabling it to change shape readily without fracture and to conform accurately to the interior of any containing vessel. This condition is attributed to lessened cohesive force, which enables the molecules to slide readily over each other without actual rupture. The molecules are somewhat more separated and vibrate with much greater activity. Yet the volume and specific gravity are usually not very different in the two phases, the liquid having a slightly greater specific weight than the solid.

The change from the solid to the liquid state of a uniform substance always takes place at a constant temperature, which is known as the **melting point.** In addition to the sensible heat which it is necessary for the substance to absorb before it can become liquid, an amount of heat peculiar to each substance is rendered latent, and is known as the **latent heat of fusion.** This heat does not appear as an increase of temperature of the substance, but is used up in separating the molecules and otherwise maintaining the state of liquidity.

Gases.—The gaseous state of matter usually results from the further application of heat to the liquid phase. A limited quantity of vapor may usually be induced by simply decreasing the pressure without change of temperature. The vaporous state is characterized by extreme tenuity, great volume and low specific gravity. The cohesive force of the substance appears to be entirely destroyed and separation may occur through its own expansive energy. Under ordinary conditions, vapor may form on the surface of a liquid until its density counterbalances the tendency to form vapor. This tendency is known as the **vapor pressure** of the liquid. The vapor pressure increases rapidly with the temperature of the liquid and when it exceeds the pressure of the superimposed atmosphere, vapor is given off rapidly and the liquid is said to "boil." This temperature is called the **boiling point.** For any given substance the boiling point is constant under given conditions. Changes of superimposed pressure, however, affect this temperature, depressing it when the pressure is decreased, and raising it when the pressure is increased. For example, in the case of water,

At sea level, under an atmos pres of 30 in. Mer, it boils at 212° F.
At 1,800 ft. elevation, under an atmos. pres. of 28 in. Mer., it boils at 208° F.
At 4,700 ft. elevation, under an atmos. pres. of 25 in Mer, it boils at 203° F.
At 10,600 ft. elevation, under an atmos. pres. of 20 in Mer., it boils at 192° F.

As in the case of fusion, the change of state from liquid to vapor is accompanied by an absorption of sensible heat which raises the temperature of the substance, and also of latent heat which supplies the energy that effects the change of state. The latter is known as the **latent heat of vaporization.** In putting the substance through the reverse changes the latent heats are again converted into sensible heat and radiated by the substance.

SPECIFIC HEATS OF CERTAIN SUBSTANCES.

Substance.	Temperatures. Deg. F.	Quantity of heat needed to raise 1 pound from 0° to t° F.	Quantity of heat needed to raise 1 pound from t_2—t_1° F.
Water	32 to 212	$1.0t + 0.00015t^2$	$1.0(t-t_1) + 0.00015(t^2-t_1^2)$
Steam	212 to 3,600	$0.42t + 0.000185t^2$	$0.42(t-t_1) + 0.000185(t^2-t_1^2)$
CO	Up to 3,600	$0.2405t + 0.00002143t^2$	$0.2405(t-t_1) + 0.00002143(t^2-t_1^2)$
CO_2	Up to 3,600	$0.1870t + 0.000111t^2$	$0.1870(t-t_1) + 0.000111(t^2-t_1^2)$
O_2	Up to 3,600	$0.2104t + 0.00001875t^2$	$0.2104(t-t_1) + 0.00001873(t^2-t_1^2)$
N_2	Up to 3,600	$0.2405t + 0.00002143t^2$	$0.2405(t-t_1) + 0.00002143(t^2-t_1^2)$
H_2	Up to 3,600	$3.367t + 0.0003t^2$	$3.367(t-t_1) + 0.0003(t^2-t_1^2)$
Air	Up to 3,600	$0.2335t + 0.0000208t^2$	$0.2335(t-t_1) + 0.0000208(t^2-t_1^2)$
MgO	Up to 3,600	$0.242t + 0.000016t^2$	$0.242(t-t_1) + 0.000016(t^2-t_1^2)$
Al_2O_3	Up to 3,600	$0.208t + 0.0000876t^2$	$0.208(t-t_1) + 0.0000876(t^2-t_1^2)$
SiO_2	Up to 3,600	$0.1833t + 0.000077t^2$	$0.1833(t-t_1) + 0.000077(t^2-t_1^2)$
CaO	Up to 3,600	$0.1715t + 0.00007t^2$	$0.1715(t-t_1) + 0.00007(t^2-t_1^2)$
Fe_2O_3	Up to 3,600	$0.1456t + 0.000188t^2$	$0.1456(t-t_1) + 0.000188(t^2-t_1^2)$

Richards, "Metallurgical Calculations," pp. 60-98 inc.

The coefficients of thermal capacity per cubic foot of the above gases may be found, if desired, by dividing the above coefficients per pound by the number of cubic feet which 1 pound of the gas occupies at 32 degrees F.

THERMAL PROPERTIES OF CERTAIN METALS.

Metal.	Melting point. Deg. F.	Latent heat of fusion B. T. U.	Boiling point. Deg. F.	Latent heat of vaporization. B. T. U.
Sodium	206	57	1,368	1,827
Magnesium	1,382	104	1,980	2,367
Aluminum	1,157	180	4,172	4,100
Silicon	2,606	230	5,072	4,540
Phosphorus	111	9	548	185
Sulphur	237	17	832	130
Calcium	1,436	95		
Manganese	3,000	132	Not readily volatile.	
Iron	2,000	126		

HEAT OF PHYSICAL CHANGES.

From the preceding discussion it is evident that in causing matter to change from a cold solid to a superheated vapor through the application of heat, there will ordinarily be five distinct quantities of heat required, as follows:

(1) Heat needed to raise the solid phase to the melting point.

(2) Heat rendered latent in effecting the change of state to liquid.

(3) Heat needed to raise the liquid to the boiling point.

(4) Heat rendered latent in effecting the change to vapor.

(5) Heat needed to superheat the vapor.

As already stated, the latent heat is the quantity of heat hypothecated by the substance to effect the change of state from one phase to another. The latent heat differs for different substances, and for any given substance the heat of fusion differs from that of vaporization. Usually they are both related to the melting point and boiling point respectively. The latent heat of fusion of elements per pound corresponds closely to the formula,

$$L.H_f = \frac{2.1 \times T(mpt).}{Atomic\ wt.}$$

where $T(mpt) =$ the absolute temperature F. of the melting point. The latent heat of vaporization corresponds closely to the formula,

Richards, Metal. Cal pp. 58 59.

$$L.H_v = \frac{23 \times T(bpt)}{Atomic\ wt.}$$

where $T(bpt) =$ the absolute temperature F. of the boiling point.

The quantity of heat needed to raise the temperature of the substance to that of the next phase depends upon its thermal capacity, which also varies for each substance and for each phase of the substance. The coefficient of thermal capacity is commonly called **"Specific Heat,"** and the method of its measurement is based upon comparison with the thermal capacity of water. The Engineers' unit of heat measurement is the **British Thermal Unit,** usually abbreviated " B. T. U." One B. T. U. is the quantity of heat required to raise one pound of water 1 degree F. at its greatest density, namely 39 degrees F. The specific heats of other substances also are the quantities of heat, measured in British Thermal Units, which are necessary to raise their temperatures 1 degree F. As the thermal capacity of water is greater than that of most substances, it follows that if its specific heat is taken as unity, those of most other substances will be expressed by fractions. The coefficients of the thermal capacity of most substances are not constant, but increase slightly as the temperature of the substances rises.

The specific heats of the elements bear a definite relation to their atomic weights. It was observed by Dulong and Petit in 1818 that the product of the atomic weight and specific heats of the elements was fairly constant at about 6.4, known as the atomic heat. It is supposed, therefore, that equal weights of all atoms have about the same capacity for heat. It follows, consequently, that an element with a low atomic weight will have a high specific heat and vice versa. This law furnishes a means of calculating specific heats from atomic weights.

The atoms appear to retain their original atomic heats even when in combination, and hence the molecular heats of compounds are equal to the sums of the atomic heats of elements forming them. Molecular heats of similar salts are virtually the same, and in simple compounds are usually approximately six times the number of atoms involved.

MOISTURE IN THE ATMOSPHERE.

All liquids and some solids give off vapor at all times, the quantity of which is dependent upon the temperature and the quantity of vapor present. The presence of other vapors or gases has no effect, except when they influence the temperature. The only pressure which appears to check vaporization is that exerted by the vapor itself. The evaporation of water is a case in point. It is customary to consider that the water vapor is taken up and held in suspension by the air, and when all of the vapor has formed that can exist at a given temperature we are accustomed to say that the air is saturated with vapor. It would be more nearly correct, however, to say that the space is saturated with vapor, because the quantity of vapor is in no way governed by the presence of air except in so far as the air influences the temperature of the space. The same amount of vapor would exist at the

given temperature if the air were wholly absent. The capacity of
space for water vapor increases rapidly as the temperature rises,
and the vapor pressure of the liquid follows closely. The change
is very slow at low temperatures, but after the temperature rises
above the freezing point the power to vaporize increases rapidly.
This tendency is illustrated graphically by curves (1) and (2) on
the chart.

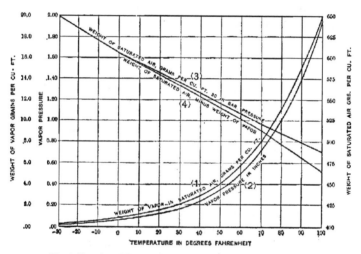

Chart Showing Effect of Temperature on Moisture in Air.

When water vapor forms in the presence of air it permeates
the air and for the time acts as an integral part of it. Water
vapor has only 63 per cent. of the weight of an equal volume of
dry air at the same temperature. Hence a cubic foot of moist air
is lighter than the same volume of air would be if dried. When
dry air takes up water vapor, however, its weight is increased by
just the weight of moisture added. At the same time, its expansive force and consequently its volume is increased, so that the
net result is less weight per cubic foot than when the moisture is
absent. If, however, the moisture could be removed from the air
without change of volume, the air so desiccated would be lighter.
This fact is shown very clearly by curves (3) and (4) of the accompanying chart. The thin wedge between the two lines (3)

and (4) shows the relative weight of the moisture and that of the saturated air.

The vaporization of water, which is so dependent upon high temperature, tends to neutralize its own activity through the cooling effect of the action itself. The heat absorbed during vaporization is 966 B. T. U. per pound, of which 894 B. T. U. becomes latent and 72 B. T. U. is used up in expanding the vapor against the pressure of the atmosphere. This quantity of heat is sufficient to raise the temperature of more than 5 pounds of ice water to the boiling point.

FIXED GASES.

While all vapors may act as perfect gases when at temperatures well above the boiling point, it is only those which do not have to be induced by elevated temperatures but whose boiling points lie well below ordinary atmospheric temperatures, that are said to be permanent or **fixed** gases. Gases are usually transparent, colorless bodies of such tenuity as to offer but little resistance to motion. They are very light yet have easily measurable weights and a given volume of a given gas at a given temperature will always weigh the same. Air, for example, is a mixture of two permanent gases and consequently is a permanent gas also. At 62 degrees F. the weight of 1 cubic foot of air at sea level is 0.0761 pounds, and it takes 13.14 cubic feet of air to weigh 1 pound avoirdupois at that temperature.

Pressure of Atmosphere.—The mass of air which surrounds the earth exerts a pressure due to its weight, equal to 14.7 pounds at sea level for every square inch of the earth's surface. This is commonly called **atmospheric pressure,** and is usually measured by means of a barometer in which a column of mercury is arranged to just balance the pressure of the air. At sea level it usually takes from 29 to 30 inches of mercury, according to weather conditions, to balance the air, which corresponds to a weight of 14.2 to 14.7 pounds atmospheric pressure. At greater altitudes the superincumbent mass of air is less and the pressure is correspondingly decreased. At 5000 feet elevation the pressure is but little over 12 pounds and at 10,000 feet the pressure falls below 10 pounds.

Weight of Gases.—The following table gives the number of cubic feet needed at sea level to make 1 pound at 62 degrees F. of the more usual gases, and also the weight in ounces per cubic foot at 32 degrees F.

At 62° F., 13 14 cu. ft. = 1 lb. air, and 1 cu ft = 1 29 oz at 32° F.
At 62° F , 11 89 cu ft. = 1 lb O_2, and 1 cu ft. = 1.44 oz at 32° F.
At 62° F , 13 53 cu. ft = 1 lb N_2, and 1 cu. ft = 1 26 oz at 32° F.
At 62° F., 189 70 cu ft = 1 lb H_2, and 1 cu. ft = 0 09 oz. at 32° F.
At 62° F , 13 55 cu ft = 1 lb CO, and 1 cu. ft = 1 26 oz at 32° F.
At 62° F , 8 60 cu ft = 1 lb CO_2, and 1 cu. ft = 1 98 oz. at 32° F.
At 62° F , 23 32 cu ft = 1 lb CH_4, and 1 cu ft = 0 73 oz at 32° F.
At 62° F , 13 46 cu. ft. = 1 lb. C_2H_4, and 1 cu ft. = 1.26 oz at 32° F.

The ratio of weight of equal volumes of any gas compared to that of air, is its specific gravity.

LAWS OF GASES.

Although the fixed gases may have very different chemical compositions, yet in physical behavior they are much alike and are all governed by a distinct set of laws which do not necessarily apply to other forms of matter. These laws define the interrelations of temperature, pressure and volume and are known as the **laws of gases.**

Law of Avogadro.—Avogadro's hypothesis that equal volumes of all gases at a given temperature and a given pressure contain the same number of molecules, implies that the molecules of all gases have the same volume. The accuracy of this statement is sufficiently well borne out by experience to justify its application. The molecules of a gas are evidently not restrained by any internal cohesion and are free to move about among each other with varying degrees of activity. Their activity is increased by increased temperature of the gas. The pressure exerted by a gas upon an enclosed space is supposed to be the manifestation of the continual bombardment of the enclosing walls by the rapidly moving molecules. The pressure naturally depends, therefore, upon both the temperature and the density of the gas. The volume, as already stated, depends upon the pressure and the temperature.

Law of Boyle.—It was observed by Boyle in 1660 that the volumes of perfect gases always vary inversely as the pressure exerted upon them, that is, a given weight of gas, if subjected to double pressure will be forced into one-half the space previously occupied In other words for a given weight of gas the product of its volume, multiplied by its pressure at any time is always constant. This relationship is expressed by the formula,

$$pv = C$$

which is perfectly true for moderate pressures. At very high pressures the law fails, as the contraction is always less than the

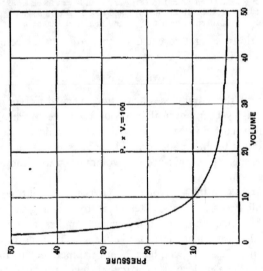

Diagram Illustrating Boyle's Law.

amount expected from the pressure exerted. This is due to the passage of a portion of the gas into an incompressible condition whose resistance to pressure is similar to that of a liquid. The quantity of the incompressible portion increases with the pressure. If the compressible portion is called v, and the incompressible portion b and the total volume of gas is called V, then we have

$$V = v + b, \text{ and } v = V - b.$$

Substituting this value of v in the formula, $pv = C$ gives the expression,

$$p(V - b) = C,$$

which is equally true for all pressures, since at low pressures b is absent and $V = v$. The incompressible portion, designated as b, resists pressure like a liquid, but does not assume the form of a liquid, however great the pressure, until it is cooled below a certain critical temperature which differs for different gases. For the more usual gases these critical temperatures have been determined to be as follows,

$CO_2 =$ .... 88°F. $N_2 =$ ....—231°F.
$O_2 =$ ....—180°F. $H_2 =$ ....—390°F.

Law of Gay-Lussac.—It was observed in 1801 by Gay-Lussac that the volumes of perfect gases, which are allowed to expand freely, vary directly as their temperatures. Experiments show

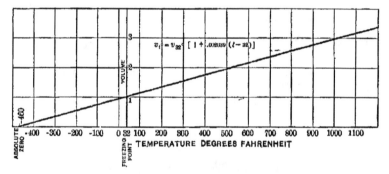

Diagram Illustrating Gay-Lussac's Law.

that for every degree Fahrenheit increase in its temperature, a perfect gas will increase each unit of the volume which it has at 32 degrees F. by an amount equal to 0.002039. This rate of increase will double the volume of the gas in 490.5 degrees F. By reverse reasoning, it might be concluded that at 490.5 degrees below 32 degrees or —458.5 degrees the gas would contract to a point and have no volume. For this reason —458.5 degrees F. is designated the **absolute zero** of temperature. The volumes of gases evidently will be proportional to their absolute temperatures. The volume of a gas at any temperature may be calculated if its volume at any other temperature is known, by means of the formula,

$$Vt = V_{32}° [1 + a (t - 32°)]$$
where
t = temperature in degrees F.
V = the volume at any given temperature.
V_{32} = volume at 32° F.
a = 0.002039.

It is only necessary to reduce the known volume to its value at 32 degrees and substitute this value in the equation for the unknown volume, thus, for example:

Let it be required to find the volume which one cubic foot of air at 60 degrees F. will occupy at 1200 degrees F.

$$V = V_{32} [1 + a (t - 32)]$$
whence,
$$V_{32} = \frac{Vt}{1 + a (t - 32)}$$
and
$$V_{32} = \frac{1}{1 + (0.002039 \times 28)} = 0.946 \text{ cubic feet.}$$

Substituting this value for V_{32} in the formula for the volume at 1200 degrees, we have,

$$V_{1200} = 0.946 [1 + (0.002039 \times 1168)] = 3.199 \text{ cubic feet,}$$
which is the volume at 1,200 degrees F.

The same result may be attained more simply by considering the volumes proportional to the absolute temperature, thus:

$$(60 + 458.5) : 1 = (1,200 + 458.5) : x,$$
whence
$$x = 3.199$$

Effusion of Gases.—The law of Effusion, announced by Graham and Bunsen, states that the rate of passage of a given volume of different gases through a porous wall or a small opening is inversely proportional to the square root of the density of the gases. The rate of transmission is, therefore, a measure of density. As the density of a given gas varies with its temperature, the rate of effusion becomes a measure of any change of temperature in the gas. This principle forms the basis of the Uehling and Steinbart pyrometer for measuring air blast temperatures.

HEAT.

Heat was once considered to be a substance which could be added or subtracted at will, but it is now known to be only a condition of matter. The present theory of heat is dynamical or mechanical in its conception. It supposes that heat is the outward manifestation of an increased activity of motion of the molecules of the substance. Heat and cold are merely relative terms, and are but manifestations of changes in intermolecular activity. Decreased activity accompanies increased cold. The theoretical minimum of temperature would represent absolute molecular quiescence and is the absolute zero of temperature.

FUEL.

Certain substances have the property of evolving heat when burned. Many such substances serve as fuels. The quantity of heat evolved by a given weight of combustible and the temperature attained by the process are not necessarily proportional, and the ratio varies in different combustibles. The quantity of heat evolved by a unit quantity of a combustible is termed the **Calorific Power** of the substance, and is measured in heat units. The temperature attained by the combustion of the substance is measured in degrees and is known as its **Calorific Intensity.** It depends not only upon the quantity of heat evolved but also upon the rate of combustion and the capacity of the products of combustion to absorb the heat. For example, the oxidation of wood during slow decay releases the same number of heat units as if the wood were cast into a fire and burned, but the calorific intensity attained in the two cases is very different. The calorific intensity may be calculated if the above factors are known, or it may be measured directly by thermometers, pyrometers or any device which is affected by changes of temperature.

Determination of Calorific Power.—The calorific power of a substance is determined by means of a device known as a calorimeter, in which the heat developed by combustion is absorbed by some suitable medium and its quantity measured. The usual absorbing medium is water. If the rise in temperature is observed, and the weight of water is known, the quantity of heat may be calculated. The amount of heat which raises 1 pound of water 1 degree F. is termed, as we have said, a British Thermal Unit, B. T. U. The calorific power of many substances has been determined by calorific measurements Some of the most important are as follows.

```
1 pound C to CO....  .....   ..  . .  4,450 B. T. U.
1 pound C to CO₂ .. ... .  . .  14,550 B. T. U.
1 pound CO to CO₂  ...............  4,325 B. T. U.
1 pound H to H₂O ..  ....   .... .62,000 B. T. U
1 pound H to steam .   ...... .....51,700 B. T. U.
```

Determination of Calorific Intensity.—Since the calorific intensity of a given fuel is the ratio between its power to develop heat and the power of the resulting products of combustion to absorb the heat produced, it follows that it can be found by simple calculation, by dividing the calorific power by the weight of the products of combustion, multiplied by their Specific Heats. For example, the calorific intensity of carbon when burned in an atmosphere of oxygen alone may be found as follows, by means of the formula for the Specific Heat of CO_2 at temperatures above 3600 degrees F.:

$$0.41t + 0.0000341t^2 = \text{heat of combustion}$$
$$C + O_2 = CO_2,$$

by which 14,550 B. T. U. are developed per pound of carbon, and the weight of the combustion product CO_2, equals $\frac{44}{12}$ or 3.667 pounds. Whence,

$$3.667 \ (0.41t + 0.000034t^2) = 14,550,$$

or,

$$0.00012468t^2 + 1.50347t = 14,550.$$

Completing the square, by adding 4529 to each side of the equation, gives

$$0.00012468t^2 + 1.50347t + 4,529 = 19,079.$$

By taking the square root we have,

$$0.01117t + 67.3 = 138.1,$$

whence,

$$0.01117t = 70.8,$$

and,

$$t = 6,338 \text{ degrees F.}$$

When, however, the combustion of carbon takes place in oxygen of the air with its accompanying nitrogen, the products of combustion are increased by just that quantity of nitrogen, which, in turn, absorbs its share of heat developed and the resulting temperature is correspondingly lower. By means of the formulas, for temperatures below 3600 degrees F.,

$$0.187t + 0.0001111t^2 = \text{heat developed, and}$$
$$0.2405t + 0.00002143t^2 = \text{heat developed,}$$

for the specific heats of CO_2 and N_2, respectively, the theoretical temperature may be calculated. Assuming no excess of air to be used, then,

$$C + O_2 + N_2 = CO_2 + N_2,$$

develops 14,550 B. T. U. per pound of carbon and the products

of combustion equal $\frac{44}{12} = 3.667$ pounds CO and $\frac{105}{12} = 8.75$

pounds N_2, whence

$$\left. \begin{array}{l} 3.667\ (0.187t + 0.000111t^2) \\ 8.75\ (0.2405t + 0.00002143t^2) \end{array} \right\} = 14,550.$$

or,

$$\begin{array}{r} 0.0004070t^2 + 0.6857t \\ 0.0001875t^2 + 2.1044t \\ \hline 0.0005945t^2 + 2.7901t = 14,550. \end{array}$$

Completing the square gives,

$$0.0005945t^2 + 2.7901t + 3,272 = 17,822.$$

The square root equals,

$$0.0244t + 57.2 = 133.5$$

Whence,

$$0.0244t = 76.3,$$

and,

$$t = 3,127 \text{ degrees F}.$$

The theoretical temperature of the combustion of carbon to the condition of CO in the presence of air may be shown as follows:

$$C_2 + O_2 + N_2 = 2CO + N_2,$$

develops 4450 B. T. U. per pound of carbon and the products of combustion are 2.333 pounds CO and 4.375 pounds N_2. As the heat capacities of CO and N_2 are identical, we have

$$6.708\ (0.2405t + 0.00002143t^2) = 4,450,$$

or,

$$0.00014375t^2 + 1.61327t = 4,450.$$

Completing the square gives,

$$0.00014375t^2 + 1.61327t + 4,637 = 9,087,$$

the square root of which is,

$$0.01199t + 68.1 = 95.3.$$

Whence,

$$t = 2,270 \text{ degrees F}.$$

The theoretical temperature of CO burned in air is somewhat higher than that accompanying its formation.

$$2CO + O_2 + N_2 = 2CO_2 + N_2,$$

whence it appears that 4325 B. T. U. are developed per pound of

CO and the products of combustion equal $\frac{88}{56}$ or 1.5714 CO_2, and

$\frac{105}{56}$ or 1.875 lbs. N_2, whence,

$$\left.\begin{array}{l} 1\,5714\,(0\,187t + 0\,000111\,t^2) \\ 1\,8750\,(0\,2405t + 0\,00002143\,t^2) \end{array}\right\} = 4,325,$$

and,

$$0\,29385t + 0\,0001744\,t^2$$
$$0\,45095 + 0.0000402\,t^2$$

$$\overline{0.74480t + 0\,0002146\,t^2 = 4,325.}$$

Completing the square gives,

$$0.0002146\,t^2 + 0\,7448t + 645 = 4,970.$$

Taking the square root gives,

$$0\,01465t + 25\,4 = 70\,5.$$

$$t = 3,080 \text{ degrees F}$$

In the case of hydrogen the product of the combustion is water. At the temperature of the reaction the water exists in the form of steam, which is a vapor and not a fixed gas such as CO, CO_2, and N_2. The vaporization of the water renders latent a large amount of the heat developed by the combustion, which is not given out unless the products of combustion are cooled below 212 degrees F. As this rarely occurs in ordinary cases of combustion, it is best to deduct the latent heat of vaporization of steam from the total heat developed and to treat it as if never produced. It amounts to about 10,300 B. T. U. per pound of hydrogen, leaving a net heat development of 51,700 B. T. U.

The calorific intensity of hydrogen burned to the condition of steam in the presence of an exact sufficiency of air may be found as follows:

$$2H_2 + O_2 + N_2 = 2H_2O + N_2,$$

by which 51,700 B. T. U. are developed per pound of hydrogen and the products of combustion are $\dfrac{36}{4}$ or 9 0 pounds steam and $\dfrac{105}{4}$ or 26.25 pounds N_2.

$$\left.\begin{array}{l} 9\,0\,(0\,42t + 0\,000185\,t^2) \\ 26.25\,(0\,2405t + 0\,00002143\,t^2) \end{array}\right\} = 51,700.$$

$$0.00166530\,t^2 + 3\,780t$$
$$0\,0005625\,t^2 + 6\,313t$$

$$\overline{0\,0022275\,t^2 + 10\,093t = 51,700.}$$

Completing the square gives

$$0.0022275\,t^2 + 10\,093t + 11,449 = 63,149.$$

Whence,

$$0\,0472t + 107 = 251,$$

and,

$$t = 3,050 \text{ degrees F.}$$

It should be observed, however, that these results presuppose that there is no excess of air over that theoretically necessary and that all of the heat developed passes into the products of combustion. These conditions are never attained in reality, as there is usually an excess or deficiency of air and much heat is absorbed by the walls of the furnace and lost by radiation and conduction. Consequently, actual temperatures are always lower than theoretical results.

APPENDIX II.

COST SHEET OF A SMALL EASTERN FURNACE DURING THE FUEL STRINGENCY IN 1902.

1902.	July.	August.	September.	October.	November.
Pig made...................	3,605	4,173	3,734	2,835	3,174
Ores	$7.219	$6 677	$7.273	$6.704	$6 884
Coke	6 637	5.083	6.657	8.898	7.983
Limestone	.701	.800	.790	.846	.982
Soft Coal..................	.066	.057	.071	.152	.116
Supplies	.096	.132	.165	.222	.140
Superintendent	.069	.020	.016	.021	.016
Pay roll..................	1 299	1.039	1.153	1,398	1 156
Labor (unloading stock)......	.300	.336	.194	.285	.234
Master mechanic............	.010	.009	.010	.013	.009
Materials for repairs.........	.009	.039	.056	.048	.045
Labor for repairs...........	.123	1.138	.160	.270	.142
Timekeepers and cost clerks...	.011	.010	.010	.017	.012
Laboratory	036	.030	.033	.047	.036
Reserve fund...............	.050	.050	.050	.050	.050
Relining fund..............	.100	.100	.100	.100	.100
Insurance	.016	.013	.015	.020	.016
Totals................	$16.742	$15.442	$16.753	$19.190	$17.924
Less by-products............	...	.276	.170	.007	.184
Net cost...............	$16 742	$15.166	$16 583	$19.183	$17.740
Cost, excluding ore, fuel and flux	2.185	1.697	1.863	2 635	1.888

INDEX.

www.ingramcontent.com/pod-product-compliance
Lightning Source LLC
LaVergne TN
LVHW012206040326
832903LV00003B/145